J. C. Polkinghorne

The Quantum World

PENGUIN BOOKS

PENGUIN BOOKS

Published by the Penguin Group
27 Wrights Lane, London W8 5TZ, England
Viking Penguin Inc., 40 West 23rd Street, New York, New York 10010, USA
Penguin Books Australia Ltd, Ringwood, Victoria, Australia
Penguin Books Canada Ltd, 2801 John Street, Markham, Ontario, Canada L3R 1B4
Penguin Books (NZ) Ltd, 182–190 Wairau Road, Auckland 10, New Zealand

Penguin Books Ltd, Registered Offices: Harmondsworth, Middlesex, England

First published by Longman 1984
Published in Pelican Books 1986
Reprinted in Penguin Books 1990
10 9 8 7 6 5 4 3 2 1

Made and printed in Great Britain by
Richard Clay Ltd, Bungay, Suffolk
Typeset in Times

Penguin Books

The Quantum World

J. C. Polkinghorne spent the greater part of his working life as a theoretical elementary particle physicist and during the period 1968 and 1979 was Professor of Mathematical Physics in the University of Cambridge. He resigned his Chair in 1979 in order to train for the ordained ministry of the Church of England. Dr Polkinghorne is a Fellow of the Royal Society, Fellow, Dean and Chaplain of Trinity Hall, Cambridge, and Honorary Professor of Theoretical Physics at the University of Kent. His most recent books include *One World* (1986) and *Science and Creation* (1988).

To my former colleagues in the Department
of Applied Mathematics and Theoretical Physics,
University of Cambridge

Contents

Preface

The physics of the twentieth century is divided from all that which
came before by two great discoveries which have transformed our
view of the natural world. One is Einstein's theory of special
relativity; the other is quantum mechanics. There is no doubt in my
mind that quantum mechanics is much the more revolutionary of
the two. Although Einstein profoundly changed our understanding
of the nature of time and the meaning of simultaneity, there is a
sense in which his work is the last great flowering of the classic tradi-
tion in physics. It preserved the clarity of description and the
inexorable determinism which had been the hallmarks of mechanics
since Newton. Quantum mechanics, on the other hand, abolished
clear-cut trajectories and introduced a probabilistic fitfulness into
nature. The resulting elusive quality bestowed on physical reality
has been the subject of much confident assertion. 'Modern physics'
has been called in aid of all sorts of mutually inconsistent
philosophical positions. To tell the truth, even the professionals
have not been immune from confusion. Unresolved matters of
disputed interpretation remain in quantum theory even after almost
sixty years of successful exploitation. This book is not written to
convey hard and fast answers to all such questions. Its purpose is to
try to indicate what is agreed and to circumscribe the areas of debate
in which men of experience and ability still take differing views.

Quantum theory is arguably the great cultural achievement of our
century. It is too important to be the preserve and pleasure of professionals alone. I have tried to write in a way accessible to any person of reasonable intelligence and persistence who is prepared to
take the trouble to look into the matter. Accordingly I have eschew-
ed the use of mathematics beyond the ability to read a simple equation (though I have provided some meatier fare in an Appendix for
those with stronger stomachs). To do so has only made things easier
in one sense. In another sense I have made things harder by denying

myself the use of that natural language which is perfectly adapted to the discussion of such matters. This has to be compensated for by a willingness to simplify at times. I believe that what I write is accurate for the purposes intended but I do not suppose that a professional could not sometimes tighten up its expression, even if his pedantry would be likely to war against the general accessibility of the argument.

A greater hazard is provided by the fact that one can scarcely write about such matters of import for an understanding of physical reality without venturing at times to make remarks of a kind which might charitably be termed 'philosophical'. I am aware that my offerings of this sort can scarcely hope to be viewed by a metaphysician as other than jejune. I am not wholly apologetic, however, as I feel that when leather is under discussion cobblers deserve a hearing. I write, not as a philosopher of science, but as a specimen physicist. To change the metaphor, I am a dinosaur and not a palaeontologist. I suppose it possible that a live brontosaur might have a thing or two to tell the students of the fossil record.

A word is necessary about the way I have decided to approach the subject. After what I trust is a gentle historical introduction, I have chosen in Chapter 3 to take the plunge into deep waters. An account, inescapably demanding if mathematically innocent, is given of the principles of quantum mechanics as the professionals use them every day. Only after that do I deal with topics like the uncertainty principle, the two-slits experiment, Schrödinger's cat and the like, which others might have chosen to use as the means of dabbling their readers' toes in the shallow end. My defence of this procedure is that the initial effort required is more than adequately compensated for by the enhanced insight with which the reader will then be able to approach these subsequent illustrations of the odd character of the quantum world. Understanding, rather than mere intellectual titillation, is what we are aiming for and I believe that this is best attained by tackling the difficult basic material first. This book may be 'physics without calculus' but it is not intended to be just a gee-whiz tour of the Strange But True.

I am grateful to Dr J. S. Bell, FRS., Professor T. W. B. Kibble, FRS., and to my son Michael, for reading the manuscript and making useful suggestions. I am also grateful to Dr Michael Rodgers for his skill as an editor and for his help in improving the manuscript.

Between the idea
And the reality
Between the motion
And the act
Falls the Shadow.

T.S. Eliot, *The Hollow Men*

Chapter 1

Perplexities

A layman venturing into the quantum world no doubt expects to encounter some fairly strange phenomena. He is prepared for the paradoxical. Yet the greatest paradox of all is likely to escape his attention unless he has a candid professional friend to point it out to him. It is simply this. Quantum theory is both stupendously successful as an account of the small-scale structure of the world and it is also the subject of unresolved debate and dispute about its interpretation. That sounds rather like being shown an impressively beautiful palace and being told that no one is quite sure whether its foundations rest on bedrock or shifting sand.

Concerning the successfulness of quantum theory there can be no dissent. From the time it reached its fully articulated form in the middle twenties of this century it has been used daily by an army of honest toilers with consistently reliable results. Originally constructed to account for atomic physics it has proved equally applicable to the behaviour of those latest candidates for the role of basic constituents of matter, the quarks and gluons. In going from atoms to quarks there is a change of scale by a factor of at least ten million. It is impressive that quantum mechanics can take that in its stride.

The problems of interpretation cluster around two issues: the nature of reality and the nature of measurement. Philosophers of science have latterly been busy explaining that science is about correlating phenomena or acquiring the power to manipulate them. They stress the theory-laden character of our pictures of the world and the extent to which scientists are said to be influenced in their thinking by the social factor of the spirit of the age. Such accounts cast doubt on whether an understanding of reality is to be conceived of as the primary goal of science or the actual nature of its achievement. These comments from the touchline may well contain points of value about the scientific game. They should not, however, cause us to neglect the observations of those who are actually players. The overwhelming impression of the participants is that they are investigating the way things are. Discovery is the name of the game. The pay-off for the rigours and *longueurs* of scientific research is the consequent gain in understanding of the way the world is constructed. Contemplating the sweep of the development of some field of science can only reinforce that feeling.

Consider, for example, our understanding of electricity and magnetism and the nature of light. In the nineteenth century, first Thomas Young demonstrated the wave character of light; then Faraday's brilliant experimental researches revealed the interlocking nature of electricity and magnetism; finally the theoretical genius of Maxwell produced an understanding of the electromagnetic field whose oscillations were identifiable with Young's light waves. It all constituted a splendid achievement. Nature, however, proved more subtle than even Maxwell had imagined. The beginning of this century produced phenomena which equally emphatically showed that light was made up of tiny particles. (It is a story which we shall tell in the following chapter.) The resulting wave/particle dilemma was resolved by Dirac in 1928 when he invented quantum field theory, a formalism which succeeds in combining waves and particles without a trace of paradox. Later developments in quantum electrodynamics (as the theory of the interaction of light and electrons is called) have led to the calculation of effects, such as the Lamb shift in hydrogen, which agree with experiment to the limits of available accuracy of a few parts per million. Can one doubt that such a tale is one of a tightening grasp of an actual reality? Of course there is an unusually strong element of corrigibility in this particular story. Quantum electrodynamics contains features completely contrary to the expectations which any nineteenth-century physicist could have entertained. Nevertheless there is also considerable continuity, with the concepts of wave and field playing vital roles throughout. The controlling element in this long development was not the ingenuity of men nor the pressure of society but the nature of the world as it was revealed to increasingly thorough investigation.

Considerations like these make scientists feel that they are right to take a philosophically realist view of the results of their researches; to suppose that they are finding out the way things are. When we are concerned with pre-quantum physics – with classical physics, as we say – that seems a particularly straightforward supposition. The analogy with the 'real' world of everyday experience is direct. In classical physics I can know both where an electron is and what it is doing. In more technical language, its position and momentum can both simultaneously be known. Such an object is not so very different from a table or a cow, concerning which I can have similar information of where they are and what they are doing. The classical electron can be conceived, so to speak, as just a midget brother of everyday things. Of course, philosophers can dispute the reality of the table and the cow too, but common sense is inclined to feel that that is a tiresomely perverse attitude to take to experience.

Heisenberg abolished such cosy picturability for quantum mechanical objects. His uncertainty principle (discussed in detail in Chapter 5) says that if I know where an electron is I have no idea of what it is doing and, conversely, if I know what it is doing I do not know where it is. The existence of such elusive objects clearly-modifies our notion of reality. One of the perplexities about the interpretation of quantum mechanics is what, if any, meaning it attaches to the reality of something as protean as an electron. In Chapter 7 we shall find that there is in quantum theory a radical inability to pin things down which goes beyond the simple considerations outlined here.

The second perplexity relates to the act of measurement. It is notorious that there is an inescapable random element in quantum mechanical measurement. Suppose that I am supplied with a sequence of electrons which have been prepared in such a way that they are all in the same state of motion. Every minute, say, I am delivered one of these standard electrons. In classical physics, if I measured each electron's position as it was delivered they would all be found to be in the same place. This is because classically they have a well-defined location at a particular instant whose specification is part of what is involved in saying that they are in the same state of motion. Quantum mechanically, however, Heisenberg will not normally allow the electrons to have a well-determined position (it will usually have to be uncertain). This reflects itself in the fact that when I actually make a measurement I shall find the electron sometimes here and sometimes over there. If I make a large number of such measurements the theory enables me to calculate the proportion of times the electron will be found 'here' and the proportion of times it will be found 'there'. That is to say, the probability of its being found 'here' can be determined. However, I am not able to predict on any particular occasion whether the electron will turn out to be 'here' rather than 'there'. Quantum mechanics puts me in the position of a canny bookmaker who can calculate the odds that a horse may win in the course of the season but not in the position of Our Newmarket Correspondent who claims to be able to forecast the outcome of a particular race.

All that is rather peculiar but it can be digested and lived with. The source of the perplexity is more subtle. Consider what is involved in measuring the position of an electron. It requires the setting up of a chain of correlated consequences linking at one end the position of the microscopic electron and at the other end the registration of the result of that particular measurement. The latter can be thought of either as something like a pointer moving across a scale to a mark labelled 'here', or ultimately as a conscious observer looking at such a pointer and saying 'By Jove, the electron's 'here' on this occasion.' Certainly I cannot perceive the electron directly. There has to be this chain of related consequence blowing-up the position of the microscopic electron into a macroscopically observable signal of its presence 'here'. The puzzle is, where along this chain does it get fixed that on this occasion the answer comes out 'here'? At one end of the chain there is a quantum mechanically uncertain electron; at the other a dependable pointer or equally reliable observer, neither of which exhibits any uncertainty in its behaviour. How do the two enmesh? In Chapter 6 we shall find that it is a matter of perplexity where along the chain the fixity sets in which determines a particular result on a particular occasion. In fact there is a range of different suggestions, none of which appears free from difficulty. The discontinuity involved in the act of measurement is the one really novel feature which sets quantum mechanics apart from all the physics which preceded it.

In what lies ahead I shall try to explain more fully the nature of these problems and the variety of the answers which have been proposed to them. Before attempting that task a short historical excursus will be necessary to explain how it all came about.

Chapter 2

How it all began

With an impeccable sense of timing the first major crack in the edifice of classical physics became apparent in 1900. Its discoverers were two English theoretical physicists, Rayleigh and Jeans. They were concerned with black body radiation.

An ordinary body exposed to radiation absorbs some of it and reflects the rest. A black body is one that perfectly absorbs, and then re-emits, all radiation falling upon it. Not only are there good approximations to such an object available in nature but also it is an ideal subject for theoretical consideration. A classic problem is posed by asking what form radiant energy takes up when confined in a container (or cavity, to use the time-hallowed language of statistical physics) whose walls are black. It can be shown that the way in which the radiant energy is distributed in equilibrium among its various frequencies (the spectrum, as we say) is independent of the details of the construction of the cavity. According to the general principles of thermodynamics it depends only on the temperature. Rayleigh and Jeans set to work to calculate this spectrum. The result was disastrous. The high frequency vibrations were so highly favoured that an infinite amount of energy would be present in them. This did not merely contradict experiment; it made no sense at all. The result was called the *ultraviolet catastrophe* – 'ultraviolet' because that means high frequencies, and 'catastrophe' for a reason too obvious to need elaboration.

About a year later a most peculiar way out of the difficulty was found by Max Planck. At the time he told his son that he thought that he had made a discovery which would prove comparable in importance with those of Newton. He was not exaggerating.

Rayleigh and Jeans, who did their calculations by different methods, had both supposed that the energy seeped in and out of the black body in a perfectly continuous way. It was the natural assumption to make. Nothing else was conceivable in the smoothly changing world of classical physics. Planck, however, had the temerity to suppose that the emission and absorption of radiant energy could take place only in the form of discrete packets. He called these bundles of energy *quanta*. By this daring stroke he succeeded in abolishing the ultraviolet catastrophe. The high frequencies were tamed and a spectrum calculated which proved to be in perfect agreement with experiment.

It was the beginning of the birth pangs of quantum theory, whose characteristic it is to replace the continuous by the discrete, the smoothly varying by the fitful. Kronecker had said that God made the integers and the rest is the work of man. It began to seem that he was right. Electromagnetic radiation oscillating v times a second turns out to be made up of a *whole number* of packets of energy, each of amount

$$hv, \qquad\qquad\qquad [1]$$

where h is Planck's celebrated constant. Just as a heap of sand is composed of individual grains, so electromagnetic radiation has a bitty structure; one never finds fractions of hv but only one, or a thousand, or a million, of these granules of energy.

The value of h is 6.63×10^{-34} joule-seconds. On the scale of everyday experience this is a very tiny quantity indeed. That is why quantum theory did not make itself felt until physics had developed to a stage where minute systems of atomic dimensions could be probed and investigated. For larger systems the bundles of energy were too numerous to be countable. A row of closely spaced dots will look like a continuous line.

At first sight it was not clear how persistent this quantum structure was. It might be that it only affected the absorption and emission processes of black bodies and did not represent an abiding individuality – like drips from a tap which merge into the mass of fluid lying in the basin. However later investigations showed that the quanta were indeed persistent. An important step in establishing this was taken by Einstein in 1905 (that amazing year in which, while working at the patent office in Berne, he also invented special relativity and explained Brownian motion). He used Planck's ideas to interpret what had proved classically inexplicable in the photoelectric effect. This concerned the way in which electrons were ejected from metals by an incident beam of light. In terms of a classical wave picture of light the electrons would all be agitated by the electromagnetic radiation, bobbing up and down like anchored buoys in an ocean swell. Whether they would break loose from their moorings and float away would depend upon the violence of the waves, the intensity of the radiation. It would not be expected to depend particularly on the frequency, the rate at which the waves rose and fell. But the unexpected was what actually happened. Below a certain critical frequency no electrons were ejected, however intense the radiation. Einstein could explain this in quantum terms. Light was behaving as a swarm of particles, a broadside of torpedoes. Only those electrons which were struck by one of these projectiles would be effected. Whether such an electron was ejected (a buoy cut loose) depended on how much energy this projectile had (was it enough to sever the mooring?). According to Planck's equation [1] the quantity of energy depended directly upon the frequency. Below the critical frequency the effect was too feeble to displace the electron (the torpedo could not wrench the buoy adrift). From these sorts of consideration it became clear that light is made up of lots of particles. It was natural to call them *photons*.

At this point we first go beyond novelty into apparent paradox, for the most outstanding achievement of nineteenth-century physics had been to establish the indubitably wave-like character of light. Those insights could not just be adandoned but no one could understand how something could be both a wave and a particle. Yet that was just what light seemed to succeed in doing. For about twenty-five years people had to live with the dilemma. Science has to cling to the available evidence even in the teeth of seeming contradiction. Then the story had a happy ending. Dirac showed that quantum theory had the remedy for its supposed ills within itself. By consistently applying quantum mechanics to Maxwell's theory of the electromagnetic field he constructed the first known specimen of a quantum field theory. This provided an example of a well-understood formalism which if interrogated in a particle-like way gave particle behaviour and if interrogated in a wave-like way gave wave behaviour. It was as if someone had asserted that it was inconceivable that a mammal should lay an egg and then a duck-billed platypus had turned up. There is no gainsaying an actual example, especially when it can be dissected and its structure understood. Since that day of Dirac's discovery the dual nature of light as wave and particle has been free of paradox for those in the know.

This has not prevented all sorts of writers, particularly those of a philosophical bent, in continuing to assert the unresolved mystery of wave/particle duality. 'We do not understand it' they say, by which they seem to mean that we cannot set it out in everyday words (though actually one can go some way towards doing so, as I tried to do in Chapter V of *The Particle Play*). This is the point at which I as a mathematical physicist begin to splutter and go red in the face. The implication seems to be that mathematics (which can be used to give a perfect articulation of the wave/particle idea) is intrinsically inferior as a mode of rational discourse to ordinary language (which cannot quite cope). It is rather like feeling that the dark companion of Sirius is somehow less certainly known because we infer its presence from its gravitational effect upon the Dog Star than it would be if we could actually see it – as if gravity were somehow less real than light. No! Mathematics is the perfect language for this sort of exercise and it shows its power by penetrating beyond the every-day dialectic of wave and particle to the synthesis of a quantum field.

The next developments involved spectroscopy, the study of the sharp coloured lines resolved by a prism when it splits up the light emitted by a heated specimen, for example when an electric discharge is passed through a gas. Spectroscopy has played a very important part in the development of quantum theory, not least because it is a branch of physics capable of attaining great accuracy in measurement and so is able to pose very precise problems for theoretical interpretation.

The first important step was made by an amateur, a Swiss schoolmaster called Balmer. A lot of people hanker after making a big scientific discovery in their spare time. Any reasonably well-known professional scientist will receive from time to time letters written by well-meaning people who indicate, usually in rather guarded terms, that they have in their possession the solution to the riddle of the universe and they just need a little help in polishing it up or propagating it. Dealing with them is rather a sad business. They do not realise that to make a discovery of some magnitude requires not only great ability and a bit of luck but also a considerable invest-ment of time and effort in mastering a demanding discipline. A tempting short cut is often suggested by numerology, fiddling around in a totally *ad hoc* way with formulae to produce a (more or less) numerical fit to some data. Most such efforts are valueless. It is good therefore to be able to record that at least one such effort scored a stunning success.

Balmer had been thinking about the spectral lines from hydrogen. The various colours of light which they manifest correspond to different frequencies of vibration of the electromagnetic field, the redder colours being of lower frequency than the bluer ones. In 1885 Balmer discovered a striking numerological relationship between the frequencies of the most prominent lines. In a slightly rewritten form due to Rydberg the frequencies v are given by the formula

$$v_n = cR\left(\frac{1}{2^2} - \frac{1}{n^2}\right),$$ [2]

where n takes the integral values 3,4,..., c is the velocity of light and R is a constant called the Rydberg. For a long time no one knew what to make of this; it appeared just a curiosity, but obviously an intriguing one.

The existence of sharp spectral lines at all was, of course, an example of discreteness (note the whole numbers in [2]) but in a form that was not obviously contradictory to classical physics. Despite the latter's continuous nature it can accommodate a certain degree of selectivity when it comes to vibrations. For example, consider a plucked string. We know that it can only oscillate at frequencies which correspond to the fundamental note and its overtones. This discreteness is due to the need to fit a *whole number* of half wavelengths along the length of the string, for which there is obviously only an enumerable number of possibilities. The classical continuity which remains for the plucked string is that each of its harmonics can be sounded as softly or as loudly as we please. Planck forbids that, for he says that the energy (equals loudness) of vibration can only come in packets of a size prescribed by [1]. But as yet he was not taken that seriously.

People therefore tried first of all to interpret [2] as corresponding to the modes of vibration of some classical system. The rather odd character of the formula called for considerable ingenuity in the enterprise. The favourite line of attack employed what was technically called the plum-pudding model. It was known that atoms were overall electrically neutral and that they contained point-like negatively charged particles, the electrons whose existence had been demonstrated by J.J. Thomson in 1897. It was, therefore, supposed that the compensating positive charge was spread out like the cakey part of a plum pudding, with the electrons embedded in it like currants. In such an environment the electrons would oscillate in ways which would depend upon the details of the positive charge distribution. The hope was to interpret [2] as the natural frequencies of the oscillation of these electrons. It was all very difficult and not very successful.

In 1911 the heavens fell in. Rutherford studied the scattering of α-particles by atoms. The principal effect would be due to their interaction with the heavy positively-charged matter in the atoms. If it was spread out in the way that the plum-pudding model supposed it would only moderately deflect the α-particles. To his surprise Rutherford found that a lot of α-particles had their motion turned through large angles; they bounced back. Getting out his old university textbook on mechanics he succeeded in showing that the behaviour observed was just what would be the case if all the positive charge were concentrated in a point-like object at the centre of the atom. He had discovered the nucleus. The plum-pudding model passed instantly away and was replaced by the solar system model of planetary electrons encircling a central positively charged nuclear 'sun'.

It was a great discovery. But it was utterly disastrous for classical physics. An electron encircling a nucleus is continuously subject to acceleration as its velocity keeps on changing direction. (Acceleration is just the rate of change of velocity.) It is an inexorable consequence of classical physics that such an accelerating electron radiates off some of its energy at a frequency corresponding to the frequency of the electron's circulation. This causes the electron to move closer to the nucleus and changes the frequency. The radiation loss continues. In other words, a classical 'solar system' atom under electromagnetic forces is nothing like the actual solar system held together by gravitational forces. The atom would be unstable, as its electrons spiralled ever closer to the nucleus, and it would emit its dying radiation in a band of frequencies with no trace of the discreteness shown by [2]. The nuclear atom dealt the *coup de grâce* to classical physics.

It was far from clear what should take its place. A valuable suggestion came from a young Dane, who after a brief stay in Cambridge had moved on to Rutherford's Manchester where the action then was. His name was Niels Bohr. He saw that if an electron could occupy any orbit round a nucleus then a spiralling collapse was inevitable. The thing to do, therefore, was to forbid it! One needed to confine the electron to certain discrete orbits only. To make life simple Bohr directed his attention solely to circular orbits. (Shades of Ptolemy and Copernicus! When in perplexity the beguiling simplicity of the circle is like a raft to a drowning man.) There was still an infinity of possible circular orbits so Bohr summoned up Planck to his aid to choose among them. Planck's constant h is measured in the same physical units as those of angular momentum, a dynamical quantity which measures the amount of rotatory motion in a system. Perhaps, therefore, angular momentum was also quantised, also came only in discrete packets. The application of such an idea would certainly select out a countable set of circles whose rotatory motions correspond precisely to the right multiple of the angular momentum quantum. In fact it turned out that the size of this quantum was given not by h but by h divided by 2π, a quantity denoted by $\hbar$ and pronounced 'aitch bar' or 'aitch slash' or often (because it really is the natural unit) just 'aitch'. (This procedure of dividing by 2π is not as forced as might appear. The ν in [1] is the number of vibrations per second. If we picture a vibration as a cycle, that is as equivalent to a rotation round a circle to arrive back at the beginning again, the angular distance travelled in the mathematician's natural units is 2π. Thus if the *angular* frequency ω replaces ν, the formula [1] changes to

$$\hbar\omega. \tag{3}$$

It turns out that this is the better way to think about Planck's condition.)

Bohr therefore supposed that the electron had to occupy a circular orbit whose angular momentum took one of the discrete values

$$n\hbar, \qquad n = 1, 2, 3, \ldots. \tag{4}$$

He could then calculate the energy corresponding to this orbit (see Appendix, A1) and it turned out to be

$$E_n = -\frac{me^4}{2\hbar^2} \cdot \frac{1}{n^2}, \tag{5}$$

where m and e are the mass and electric charge of the electron. That $1/n^2$ is very exciting when we recall Balmer's formula [2]. If an electron jumps from an orbit n to the orbit with $n = 2$ the loss of energy is

$$\frac{me^4}{2\hbar^2} \left(\frac{1}{2^2} - \frac{1}{n^2} \right). \tag{6}$$

That energy must go somewhere, presumably as radiation. Matching one discreteness with another Bohr supposed that the energy loss was radiated as a single photon. By Planck's formula in the form [3] we see that the corresponding frequency is

$$\omega_n = \frac{me^4}{2\hbar^3}\left(\frac{1}{2^2} - \frac{1}{n^2}\right). \tag{7}$$

It was a brilliant achievement. Not only had Bohr explained Balmer's mysterious formula (and other later formulae for lines corresponding to ending up in orbits with $n = 1$ and $n = 3$) but he had also enabled the value of the Rydberg constant R to be calculated in terms of the known quantities m, e, c and $\hbar$. Of course his answer for R agreed with that determined experimentally. Bohr's triumph was complete.

Nevertheless, it was achieved, not by the creation of a new theory, but rather by what amounted to an inspired piece of tinkering with classical physics. Bohr's work was full of special pleading: the use of circular orbits, the *ad hoc* imposition of the quantum condition for angular momentum. Some generalisations proved possible, but only in a piecemeal and not wholly consistent fashion. The new wine of quantum theory was soon bursting the old wineskins of classical mechanics. The Bohr atom was just a staging post on the way to the quantum world rather than the point of entry into the land itself.

One promising line of further reconnaissance involved the attempt to find a fundamental role for waves where no one had previously supposed they were relevant. We have already seen how naturally the need to fit a certain number of half wavelengths into an interval leads to discreteness and it is some way of introducing a radical discreteness into mechanics which we are looking for. This line of attack was given a great impetus when it was suggested that the wave/particle duality characteristic of light might be a universal phenomenon. In his Ph.D. dissertation the young Prince Louis de Broglie proposed a way in which waves might be associated with objects like electrons which had hitherto been thought of in purely particle terms (see Appendix, A2). His suggestion found vindication in the discovery of the phenomenon of electron diffraction. This was the exact analogue for electrons of the demonstration of the wave character of light that Thomas Young had given in 1803. The rigid separation of the world into billiard ball particles and aether waves, which was the nineteenth-century physicist's view, had completely dissolved.

It was one thing to talk in ambiguous wave/particle terms. It was quite another to find a quantitative formalism with which to calculate. There was, however, one promising possible analogy. A novice student of optics does not need to perform elaborate wave calculations to trace light through a simple system of lenses and mirrors. He just draws some rays, converging at appropriate foci for the lenses and reflected from the mirrors with equal angles of incidence and reflection. He is making use of the theory of geometrical optics which is an excellent and well-understood approximation to the wave theory, provided the wavelength of the light is small compared to the physical dimensions of the system under consideration. The rays of geometrical optics are not unlike the particle trajectories of classical mechanics. Perhaps the latter is just a short wavelength approximation to a true wave mechanics which we can attempt to unscramble by analogy with the relation that geometrical optics bears to wave optics. This programme was successfully carried through by Erwin Schrödinger and his results published early in 1926, a classic counter-example to the assertion that distinguished theoretical physicists do their best work before they are 25 (Schrödinger was 38 at the time). His work led to the celebrated equation which bears his name, an equation which can be found written down at the start of every book on quantum mechanics. Its explicit discussion requires a little bit more mathematical knowledge than our self-denying ordinance permits us to use in the body of the text (see Appendix, A3).

One of the first systems to which the Schrödinger equation was applied was the simplest atom, hydrogen. The outcome showed both how near and how far Bohr had been with his improvising stab at reality. The possible energies turned out to be given by a formula exactly like equation [5] with n taking the integer values 1,2,3,. . .. Without that result the theory would have failed, for it would not have explained Balmer's formula. The quantity n is called the principal quantum number and though it is related to angular momentum it is not identical with it as Bohr had supposed. Angular momentum does indeed come in packets of $\hbar$ and its true relation to n is that in a state whose energy is given by [5] the electron can have an angular momentum of anything from zero up to $(n - 1)$ packets of $\hbar$. Needless to say the electron is not to be considered as encircling the nucleus in a circular orbit but instead it is spread out in a way that is totally unpicturable classically. There is a unique state of lowest energy corresponding to $n = 1$. Once the atom is in this ground state, as it is called, it cannot lose any more energy. This explains the remarkable stability of atoms. It had been a puzzle that they could be knocked around in interaction with each other and yet emerge unscathed, the same as they had been at the beginning. It was now clear that this was because an atom in its ground state has nowhere else to go, unless it can be given the rather large amount of energy necessary to lift it to an excited state with n greater than 1.

In 1925, a little before Schrödinger produced wave mechanics, Werner Heisenberg, recuperating on the island of Heligoland from an attack of hay fever, had invented a theory which he called matrix mechanics. It also succeeded in bringing a desired discreteness into physics. For a little while it was not clear that wave mechanics and matrix mechanics were different expressions of the same basic physical theory, but in fact that proved to be the case. This was most clearly shown by Paul Dirac's formulation of the general principles of quantum theory, of which we shall give some account in the following chapter. A very important step was taken when Max Born made it clear that the waves of the new theory were probability waves. An electron no longer always had a specific location. One could only say of an electron encircling the hydrogen nucleus that it had a certain probability of being found here and a certain probability of being found there.

In those marvellous years 1925 – 26 fully fledged quantum theory came into being. This creative period was followed by an exploitive period in which many theoretical physicists moved in to apply the new mechanics to a host of significant problems. Speaking of that exploitive period Dirac once said to me 'It was a time when second-rate men did first-rate work.' The speaker was an undoubted first-rate man whose achievements make him the greatest British theoretical physicist since Maxwell and one of the great figures of twentieth-century culture. His observation, made in a characteristically matter of fact manner, was in no way intended as a put-down of his intellectual inheritors but simply as an indication of the fertility of that great harvest period. That is certainly how those years are enshrined in the folk memory of theoretical physicists. Whenever there is a puzzling development in physics which seems for a while to put in question our basic understanding, you will hear someone say 'I have the feeling that it's 1925 all over again.' There is a wistful note in such a comment. The speaker feels – and don't we all? – that if he had been around then he might have had the good fortune to cut a greater dash in the subject than he is succeeding in doing in present circumstances. Be that as it may, though many exciting things have happened to physics in the interim, there has not been another time since 1925 when the foundations have been shaken, the paradigm shifted, our view of the world transformed. The quantum theory created then has proved adequate for all that has so far followed it.

Chapter 3

The tools of the trade

The strange fact that the professional practitioners of quantum mechanics can calculate away without worrying about the outcome of the debates about the foundations of their subject is only possible because there is a well-defined set of rules about how to set up the problems and how to do the sums. Whatever perplexities there may be about ontology (that is, about the nature of the reality described) there is no perplexity whatever about the procedures. In this chapter I shall try to describe the quantum mechanic's tool kit with which he plies his trade.

The first notion that we require is that of a *state*. A dynamical system – a photon, an atom, what you will – is in a definite state when we know as much about its motion as physics permits. In the words of Dirac

a state of motion of a system may be defined as an undisturbed motion that is restricted by as many conditions or data as are theoretically possible without mutual interference or contradiction.

Of course, there is nothing intrinsically quantum mechanical in what has been said so far. Our definition is equally acceptable to the classical physicist. Quantum mechanical states, however, differ in two important respects from those of classical physics.

The first difference is simply that we are allowed to specify less in the way of 'conditions or data' in order to fix the state. For a classical particle we can know both its position and momentum, where it is and what it is doing. Quantum mechanically we may know either its position or its momentum but not both. This paucity of information arises from the role that uncertainty has in quantum mechanics. So far we have just asserted that this is the case. The full discussion must wait for Chapter 5 but we can begin to get an insight into how this comes about when we consider the second property of quantum mechanical states.

This says that they satisfy a *superposition principle*. The idea is a strange one and it will require some preliminaries before it can be explained. As always, the behaviour of light is a useful source of illustration.

A property which light possesses is that of polarisation. In wave terms this corresponds to the direction in which the oscillations take place. The waves are transverse, which means that the direction of oscillation has to be perpendicular to the direction of motion of the wave. If light is moving in the direction labelled z in the figure, it has two distinct possibilities of polarisation, corresponding to oscillations in the two mutually perpendicular directions x and y respectively. Those with a knowledge of simple trigonometry will recognise that oscillation in any other direction perpendicular to z, say in the direction x', could be made up from a component proportional to $\cos \alpha$ in the direction x and a component proportional to $\sin \alpha$ in the direction y.

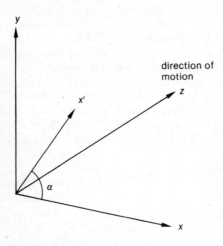

In more mundane language, what we are saying is simply that a movement to a point along x' could be executed in two steps, one along x and then another along y. These two steps are called the components of x' along x and y. The corresponding component oscillations of the wave would be, as the physicists say, 'in phase', that is the crests and troughs would occur at the same instants for both components. It is also possible to make more subtle combinations of the oscillations in the x and y directions in which the components are 'out of phase', their crests occurring at different instants. For example, so-called circularly polarised light consists of oscillations in which crests in the x direction coincide with zeroes in the y direction. Do not worry if that sounds a bit complicated. The essential idea is that all the possible states of oscillation of light moving in the z direction can be made up of judicious mixtures of the two basic oscillations in the x and y directions.

There are crystals, of substances such as tourmaline, which are sensitive to the polarisation of light. The crystal will only let through light which is polarised perpendicular to a certain axis, called the crystal's optical axis. When light which is polarised in a direction making an angle α with the optical axis falls on the crystal only a fraction $\sin^2\alpha$ of the beam is found to be transmitted. From a wave point of view this is easily understood. If the axis x of the figure is the optical axis then only the component of the light in the y direction will be let through. By trigonometry, we have already said that this component has magnitude proportional to $\sin\alpha$. Its intensity (that is, its energy content) is given by the square of this, so we expect a transmitted fraction of $\sin^2\alpha$, as is indeed the case.

That is talking wave language. However, we know that quantum mechanically we can also think of the beam as made up of photons. Suppose the intensity is lowered to a level at which only a single photon at a time encounters the crystal. Such a photon is in rather a quandary. To agree with the results it should only allow a fraction $\sin^2\alpha$ of its energy to get through. But our photon is indivisible. Either it gets through completely or it totally fails to do so, the whole hog or nothing. The only way out of the dilemma is to suppose that sometimes the photon gets through and sometimes it does not. What will happen on a particular occasion we cannot say, but we can predict that after a large number of photons have encountered the crystal a fraction $\sin^2\alpha$ will be found to have been transmitted. In other words, the best we can do is to assign a probability $\sin^2\alpha$ for the photon to get through (and, of course, a complementary probability $\cos^2\alpha$ that it does not). The radical unpredictability of individual events in quantum theory has made itself felt.

Another interesting aspect of this experiment refers to the polarisation properties of the transmitted light. It is found that its direction of polarisation has been changed on traversing the crystal from being along x' to being along y. That is, it is wholly polarised perpendicular to the optical axis of the crystal. This can be demonstrated by making the transmitted beam pass through a second crystal. If the latter's optical axis is parallel to that of the first crystal than the beam is totally transmitted, if it is perpendicular to the first optical axis none of the beam gets through. These results imply that the transmitted beam was polarised along y. From the wave point of view this is hardly a surprise, because it was just the y component of the oscillation which got through. From the particle point of view the result is less obvious. Each transmitted photon has had its polarisation state changed instantaneously from being along x' to being along y. (The alert reader may wonder what a polarisation state can mean for a *particle*. Trust me that it can be interpreted appropriately. The physically knowledgeable may be helped by the remark that it corresponds to a state of the particle's spin.)

All these particle facts can be put together in the following way: A state (the photon polarised along x') can be thought of as composed of a combination (the technical term is a superposition) of other states (the photon polarised along y, which is transmitted, and the photon polarised along x, which is not transmitted). The properties of the state so formed are related in a probabilistic way to the properties of the states out of which it is composed [the photon has a chance $\sin^2 \alpha$ of transmission (which is a certainty for polarisation along y) and a chance $\cos^2 \alpha$ of not being transmitted (which is a certainty for polarisation along x)]. This is what is meant by the superposition principle: that states can be combined in this way with a probability interpretation of the result.

It is difficult at first encounter to appreciate how great a step has been taken in adopting this principle. Perhaps another example will help. I learnt my quantum mechanics, so to speak, straight from the horse's mouth. I mean that I attended the lectures given at Cambridge by Dirac. They were based on his book *The Principles of Quantum Mechanics*, one of the intellectual classics of this century. Their style was exceptionally clear and one was carried along in the unfolding of an argument which seemed as majestically inevitable as the development of a Bach fugue. Gestures, rhetorical or otherwise, were kept to a minimum. However, early on, as he introduced the superposition principle, Dirac would break in half a piece of chalk. There was a state, he said, where the chalk was here – placing one of the pieces on his desk. There was another state where it was there – placing the second piece on the other side of the desk. According to quantum mechanics there were also states formed by the combination of these two possibilities, where the chalk would sometimes be found here and sometimes there. In other words, the superposition principle takes us straight to the heart of quantum mechanics' elusive indeterminacy.

At its mathematically most basic, superposition is just the adding together of quantities of different sorts. Suppose I walk five steps in a direction 53.1° east of north. I end up in exactly the same spot as if I had first walked four paces due east (which is one sort of displacement, an easterly sort) and then three paces due north (which is another sort of displacement, a northerly sort). I can, therefore, think of my movement as a superposition, or addition, of these easterly and northerly movements. In a similar way the quantum mechanical state of an electron can be a superposition of the state in which it is 'here' and the state in which it is 'there'. The unexpected feature of the quantum world is that the state resulting from this combination does not correspond to the electron being somewhere in the middle between 'here' and 'there' but rather to its having a certain probability to be found 'here' and a certain probability to be found 'there'.

The reader who has grasped these ideas has the root of the matter in him. The professional, however, will wish to express the notion in as general and concise a way as possible. In a word, he will want a mathematical formulation of it. It is worth the effort to convey something of how this is done but do not worry if you find the remaining paragraphs of this section rather abstract and demanding. They are only expressing more precisely what we have already said.

There is in fact a formalism to hand which perfectly expresses the idea of superposition. It is called the theory of *vector spaces*. In attempting its exposition I face the problem of a generation gap. Those of my readers who are young enough to have been exposed to 'modern mathematics' at school, and for whom it 'took', will be familiar already with the essentials of this rather sophisticated notion. For them vectors will be ordered sets of numbers which can be added together and multiplied by factors (scalars). These mathematical concepts are obviously rather well suited to represent the physical idea of superposition, which involves adding a bit of this to a bit of that. For those in this happy state the point is prac-

tically made already. They will simply need the occasional paren-
thetic remark in the course of what follows.

For the rest, the older generation and the perennially
mathematically innocent, it will be best to start by thinking of a
vector as something like an arrow directing us from one point to
another. Such displacements can be decomposed into constituent
displacements along standard directions. Our original vector is then
made up of these constituent vectors just as our quantum
mechanical states were composed of combinations of other states.

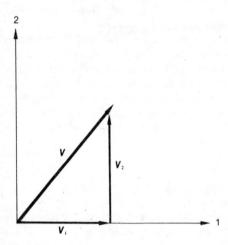

All that is a bit abstract but it can be earthed by an example. Take
the vector V of the figure. It is obviously equivalent to the pair of
displacements V_1 along the 1 direction and V_2 along the 2 direction.
So I write

$$V = V_1 + V_2. \tag{1}$$

Moreover, if a unit displacement along the direction 1 is represented
by 1, then V_1 is simply $|V_1|\,1$, where in a standard notation $|V_1|$
denotes the magnitude of the displacement V_1 (its length in other
words). Similarly V_2 can be written $|V_2|\,2$. With the conciseness of
mathematics I can therefore write

$$V = |V_1|\,1 + |V_2|\,2. \tag{2}$$

You can see that I am representing any vector V as a superposition of two standard vectors 1 and 2, with coefficients (as we say) given by the numbers $|V_1|$ and $|V_2|$. [For the modern mathematician these numbers would make up the ordered pair $(|V_1|, |V_2|)$, and our standard vectors would be $1 = (1,0)$, $2 = (0,1)$.] This property of superposibility both enshrines the pure mathematician's abstract concept of what a vector space is and it also provides the perfect way in which to mirror the nature of quantum mechanical states.

Vectors that I can represent as arrows on a sheet of paper will be two dimensional, that is to say they will be combinations of just two independent basic vectors (1 and 2 in my example, or 'across' and 'up' in words; I do not need further vectors 'down' and 'backwards' since I can regard these as 'up' and 'across' with negative coefficients. Just as $1 + (-1) = 0$, so a step 'up' and a step 'down' added together leave me where I started with zero displacement). The world of everyday experience is three dimensional and to represent it I shall need to add a third basic vector, 'out' of the paper. The mathematician can readily conceive of spaces with as many dimensions as he pleases, in which he writes

$$V = \lambda_1 1 + \lambda_2 2 + \lambda_3 3 + \dots, \qquad [3]$$

with as many terms as there are dimensions to the space. [That would correspond to the ordered n-tuple $(\lambda_1, \lambda_2, \lambda_3, \dots)$.] With a little care one can even have an infinite number of terms on the right hand side. I have written the numbers λ_i for the quantities $|V_i|$ of [2].

The choice of dimensions appropriate to a quantum mechanical problem will depend on the number of independent possibilities the system possesses. Sometimes there are just two, as when we were considering the polarisation of light. Another two-dimensional example is provided by the spin of the electron. Electrons behave 'as if' they were midget tops. I put 'as if' in quotation marks because electron spin is not a phenomenon which can be visualised quite as simply as that. It turns out that if I choose a direction and look at the electron's spin in relation to that direction then either its axis of rotation points along that direction or it is wholly in the reverse direction. That is to say, the spin is either 'up' or 'down' with respect to that direction and it can never be a bit 'sideways'. At this point I have to appeal to the trust of the reader. What I have just said shows in fact how very *unlike* a midget top an electron actually is. Such a top could certainly spin about a sideways direction but the electron is restricted to the 'up' and 'down' possibilities. It sounds rather weird, and so in a way it is, although it finds a natural expression in terms of the quantum mechanical formalism. It is the latter statement which you have to take on trust. It is not possible in a modest book like this to clarify everything. For the moment just note that electron spin provides a second example of a two-dimensional state vector space in quantum mechanics.

By contrast, in many problems the number of possibilities is unbounded and then one must use infinite-dimensional spaces to represent them. That may sound a little daunting, but it is quite a natural possibility, known to the mathematician as a *Hilbert space*. (For more detail see Appendix, A4.)

There is one further matter to confess. It would be natural to suppose that the numbers λ_i in equation [3] were ordinary real numbers, like 0.7 or -1.2. That would certainly be the case for vectors drawn on the page, which I used as my opening illustration. However quantum mechanics requires a generalisation of this in which the λ's are allowed to be complex numbers, that is to say to include i, the notorious square root of minus one. Thus a λ could be something like $1.3 - 0.4i$. If that is the mathematical straw which breaks the camel's back for you, don't worry. I had to disclose the fact but its full comprehension will not be essential in the sequel, however vital it is for anyone attempting a professional grasp of quantum theory. If on the other hand you are someone at home in the complex plane then you may be able to recognise that the argument of a complex number will provide the sort of phase relationships that we hinted at when we mentioned circularly polarised light. (A factor of i provides the phase difference of $\pi/2$ necessary to shift from a crest to a zero.)

We have made considerable progress since the beginning of this chapter. We need to go on further still but before we do so a little breather for taking stock would seem appropriate. States of motion in quantum mechanics can be superposed in a way that is analogous to the compounding of displacements in space or, more generally, to the addition of vectors in an abstract vector space. The result of such a superposition produces a state whose properties are related to those of its component parts in a probabilistic sense. That is to say, the superposition of a state with an electron 'here' and a state with an electron 'there' does not produce a state with an electron at some point in between 'here' or 'there' but a state in which the electron can be found either 'here' or 'there', with probabilities which depend (in a way I have not defined but which is capable of precise specification – see Appendix, A5) on the balance between the two states. If the state is mostly 'here' with just an admixture of 'there', then the electron will much more frequently be found 'here' than 'there'. On any particular occasion of measurement, however, we are unable to predict which possibility will be realised.

You may wonder how this rather high-falutin' talk of vector spaces is related to the simpler language of wave mechanics as it was presented in the last chapter. The abstract vectors of quantum theory correspond to the wavefunction (traditionally denoted by ψ) of Schrödinger's wave mechanics. Superposition is of the essence of waves, which can be added together to reinforce or cancel each other out.

Our description of nature is so far only half complete. In any experiment there are two steps involved. The first is the preparation of the specimen, the production of an electron, for example, in some standard state of motion. The electrons may be emitted from a hot cathode maintained at a fixed temperature and then conducted through a stable arrangement of electric and magnetic fields which imparts to them a certain amount of energy and focusses them in a particular region of space. This step is the preparatory part of the process and in its ideal form it produces an electron in one of those well-defined states of motion which we have learnt to think of as represented by a vector in a vector space. Now comes the second step, the act of measurement itself, the ascertaining where the electron is or how it is moving. Professionals speak of this as observing the value of a dynamical variable. Such *observables*, as we call them, include all those quantities which are the stock-in-trade of the theoretical physicist – position, momentum, energy and the like. We shall not have the makings of a dynamical theory until we have succeeded in finding a way to represent within our formalism these dynamical variables, these observable quantities of the theory.

If we are on the right track with the vector space idea it is natural to hope that the mathematics might contain within itself some objects ripe for interpretation as physical observables. That hope proves justified. The key notion is that of a *linear operator* on a vector space.

We need, therefore, to sally forth once more into the mathematical jungle of vector spaces. The going will be a bit tough, not because I shall use sophisticated techniques (I shall not), but because I cannot avoid producing in fairly quick succession a sequence of ideas which will be new to many of my readers. We do not want to lose any explorers of the quantum world on the way so I think I had better issue a survival kit before we start. The most useful part of such a kit will be a map indicating our destination and the way to it. Here is that map.

We are looking for two things. The first is a set of mathematical objects which will serve to represent the physically observable quantities in a similar way to that in which we used vectors to represent the physical states. These mathematical counterparts of physical acts of measurement are found in that part of the jungle where the map is marked 'here are operators'. An operator is simply something which turns one vector into another. An example would be the operation of anticlockwise rotation through 90°, which turns 'three steps east' into 'three steps north'. Actually, that definition of an operator proves to be a bit too general so that I have to narrow my choice somewhat, in a way I shall describe shortly. If you find that what I have to say about the specialisation is difficult, don't worry. Just hold onto the idea that certain operators will possess mathematical properties which make them perfectly suited to the role of representing physical observables.

So far, so good; but the results of actual measurements are always *numbers*, not abstract things like operators. If this programme is to make any sense we shall have to find a way to associate numbers with our operators. That is our second objective. It is accomplished in that part of the jungle where the map is marked 'here are eigenvalues and eigenvectors'. Once again, if you find my explanation difficult at first reading, don't despair. Just accept that there is this mathematically natural way of meeting this physical need.

The traveller who reaches this point is rewarded by an insight into why our powers of measurement in quantum mechanics are more restricted than they are in classical mechanics; why we cannot, for instance, measure both the position of an electron and its momentum. This physical consequence is seen to follow from the mathematical fact that our operators are what is called non-commutative. I explain what that means in the final paragraph of this section.

It is time for our expedition to begin. Our starting point is that an operator is simply something which, given a vector, is a means of associating another vector with it. If you like, an operator O corresponds to a rule for turning a vector V into another vector V';

$$O : V \rightarrow V' \qquad [4]$$

is a succinct way of writing it. In the world of men, robbery is an operator for it is a procedure for turning men with money in their pockets into men without, and vice versa. It transforms a state of wealth into a state of penury.

The act of passing light through a polariser also acts as an operator, in this case applied to the physical states of the photon's polarisation. We have seen that it turns a state of arbitrary polarisation into one polarised perpendicular to the crystal's optical axis. Now we can also think of the use of a polariser as being equivalent to a measurement of the polarisation, since we know that the transmitted beam has to have its polarisation in this perpendicular direction. Perhaps this observation begins to afford a clue to how it comes about that observable quantities can be associated with certain operators.

Our first degree of specialisation to make the notion useful is to restrict ourselves to linear operators. These have the properties of transforming sums into sums and multiples into multiples. Formally O is linear if, when it turns V_1 into V_1' and V_2 into V_2', then also

$$V_1 + V_2 \rightarrow V_1' + V_2', \qquad [5a]$$

$$\text{and } \lambda V_1 \rightarrow \lambda V_1'. \qquad [5b]$$

Robbery is a linear operator. It turns Jones into penniless Jones, Smith into penniless Smith, and if I rob Smith and Jones together they haven't a penny between them. That is the analogue of [5a]. If I rob identical twins I produce two similarly penniless men. That is the analogue of [5b] with $\lambda = 2$. Similarly, a polariser can be shown to act linearly. By no means all operators are linear however. The operation of squaring numbers is a counter example. It turns 1 into 1, 2 into 4, but since it turns $1 + 2 = 3$ into 9, which we can all agree is not equal to $1 + 4$, it does not satisfy [5a]. Even linear operators are too general for quantum mechanics for a reason to which we must now turn.

It is all very well to talk about observables being represented by operators but the result of an actual measurement is always just a number. Whatever abstract mathematical objects I may choose to represent energy or momentum, the consequence of actually measuring them is the concrete one of finding that I have present so many units of whatever it is. In other words, if this programme is to succeed there has to be some way of associating plain honest numbers with abstract operators. Fortunately mathematics is equal to the challenge.

Its answer can be illustrated by considering again the action of a polariser. A state which is intially polarised perpendicular to the optical axis is transmitted entirely unscathed, whilst a state which is polarised parallel to the optical axis is totally extinguished. If we represent these facts in mathematical form we would say that for the first state (let us denote it by V_{perp}) the polariser operator P has the effect:

$$P: V_{perp} \rightarrow V_{perp}, \qquad [6]$$

for that is just the mathematical way of saying that this state is unchanged by the action of the polariser. On the other hand, for the second state (V_{par})

$$P: V_{par} \rightarrow 0, \qquad [7]$$

for that is the just the mathematical statement that the polariser extinguishes this state.

These are particular examples of what the mathematicians call eigenvector relations. If an operator O turns a vector V into a multiple v of itself,

$$O: V \rightarrow vV, \qquad [8]$$

then the vector V is said to be an *eigenvector* of O with *eigenvalue v*. Equations [6] and [7] illustrate this notion for the operator P with eigenvalues $v = 1$ and $v = 0$, respectively.

This notion of eigenvalues gives numbers a natural lodgement in the theory of operators acting on vector spaces. It thus affords a way in which the results of measurement can be associated with quantum mechanical observables: the possible results of measuring an observable are just the set of eigenvalues of the corresponding operator. [The polariser is measuring polarisation perpendicular to the optical axis. If that is indeed the state of polarisation (V_{perp}) we get the expected answer 1; if that state of polarisation is wholly absent (V_{par}) we get the appropriate answer 0.] The eigenvectors must obviously correspond to special states. They are taken to correspond to those states in which the observable definitely takes that particular value. (V_{perp} and V_{par} *are* the polarisation states in question.) Thus, for example, if an electron is in a state of definite momentum p (rather than one in which its momentum might be found to take one of a possible range of values) then that state is an eigenstate of the momentum operator with eigenvalue p.

If this idea is to work it will be necessary to ensure that these eigenvalues are all real numbers. Remember that elsewhere our formalism contains complex numbers. Advantageous as that is in its proper place it would be disastrous if such numbers turned up as eigenvalues. Quantum mechanics may be peculiar but not to the extent of allowing one to obtain the square root of minus one as the result of an experiment! The mathematicians guarantee that all will be well provided we restrict ourselves to operators which satisfy the condition which they call being *hermitean*. (For a definition, see Appendix A5.) With this restriction we can be sure that the theory will make sensible predictions about the results of measurement.

We are now able to understand why our information about the states of motion is so restricted in quantum mechanics. If we could know both where a particle was and also what it was doing, it would then have to be in a state which was simultaneously an eigenstate of the position operator x and also an eigenstate of the momentum operator p. The mathematicians tell us that this would only be possible if the operators x and p were to commute. By that they mean that multiplying them together in either order would have to give the same result. Of course ordinary *numbers* do have this commutation property; 2 x 3 and 3 x 2 are both undoubtedly 6. However, for *operators* this is not usually the case. In quantum mechanics it turns out that the operators x and p do not commute, which is why there cannot be a state in which they both take definite values.

Operators which do not commute can be illustrated by considering the action of two polarisers P_1 and P_2 whose optical axes make an angle α with each other. Consider a photon which is polarised parallel to the optical axis of P_1. If it first encounters P_1 it cannot get through at all, since it is the state which is extinguished by P_1. However, if it encounters P_2 first, then there is a chance $(\sin^2\alpha)$ of transmission and, since the transmitted photon is then polarised in a new direction (perpendicular to the optical axis of P_2) there is a further chance $(\cos^2\alpha)$ that it could subsequently be transmitted by P_1. Thus encountering the polarisers in the order $(P_1$ then $P_2)$ completely blocks our initial photon, whilst encountering them in the order $(P_2$ then $P_1)$ gives it a chance of transmission. The order of encounter relates to the order in which the corresponding operators would be multiplied together in the mathematical formalism. The fact that the two results are different shows that this order matters; the operators do not commute.

It is time to take a second tea-break in our tráining as apprentice quantum mechanics. Dynamical variables, such as energy, momentum and position, have been successfully introduced into the formalism as hermitean operators acting on the vector space of the states of motion. The real numbers which result from making measurements are interpreted as the eigenvalues of the corresponding observable. Two observables which do not have the special property of commuting with each other will not be simultaneously measurable.

Once again we are saying something in general form which corresponds exactly to the particularities of wave mechanics. Schrödinger represented the momentum p by a differential operator $-i\hbar \, d/dx$ which does not commute with the position operator x. Let those who are literate in the calculus comprehend! Those who are not must forgive me my momentary lapse into modest mathematical sophistication.

What we have succeeded in doing so far is to set the scene for the quantum mechanical play. In the language of the physicist, we have given an account of the kinematics. All the scenery of states of motion is in place and the caste of observable quantities assembled. It is time for the plot to unfold. In technical language, we must add to the kinematics a dynamics, an equation of motion which will prescribe how things will evolve with time. It turns out that there are a number of equivalent ways in which this can be done. The reason for this lies in the fact that the physics depends neither on the states alone nor on the observables alone but upon their interrelation. (Experiment involves the measurement of a particular quantity in a particular state.) Thus the changing nature of physics can be attributed to the changing states, or to changing observables, or to some mixture of responsibility between the two of them. We have the choice (see Appendix, A.6). The reason that Schrödinger's and Heisenberg's versions of quantum mechanics had seemed at first sight different from each other (see p. 14) was that they had chosen to use contrasting extreme possibilities. Schrödinger made the states the sole source of change, whilst Heisenberg attributed it all to the variation of observables. In a bizarre extension of our theatrical metaphor, Schrödinger moved the scenery with the actors standing still, whilst Heisenberg left the scenery untouched and let the actors move around. This way of putting it at least has the merit of correctly suggesting that Heisenberg's procedure was closer to the way in which people were used to thinking about dynamics. Nevertheless it will be the Schrödinger point of view that will be better for us to follow.

It is enshrined in his famous equation. Whatever the risk of shock to my more mathematically allergic readers I must write this celebrated equation just once:

$$ i\hbar \frac{\partial}{\partial t} \mid \psi \rangle = H \mid \psi \rangle. \qquad [9] $$

The left hand side is $i\hbar$ times the rate of change with time of a state vector. The right hand side equates this with the effect of an operator, the *Hamiltonian*, which is simply the observable corresponding to the energy of the system under consideration. For the state vector I have used a mixed notation. The symbol ψ reminds us that it is a generalisation of Schrödinger's wavefunction. Indeed in loose quantum mechanical talk we tend to use the terms 'wavefunction' and 'state vector' interchangeably. However the proper symbol for a state vector is $\mid \rangle$. It is like half a bracket and in fact it is called a *ket vector*. (The so-called bra half is also used and written $\langle \mid$. To explain what purpose it serves would take me too far into the mathematical mysteries of dual vector spaces and scalar products – see Appendix, A4.) This bra – ket terminology was invented by Dirac. When he lectured he never made the slightest attempt to underline what had been his own (considerable) contributions to the subject. Nevertheless, a tiny smile did play round his features when he introduced bras and kets and one had the feeling that this little joke had given him as much pleasure as anything.

Whatever you can make of equation [9] there is no denying that it is a differential equation, not so very different in its way from the differential equations that Newton and Maxwell had used when they had created the fundamental basis of classical physics. To be sure there are some differences. Classical physics tends to express things in terms of second order differential equations, that is ones involving the rate of change of a rate of change, whilst [9] is first order – it just incorporates a simple rate of change. (Mathematically this is connected with the fact that classically we can specify more initially – both position and momentum – than is permitted quantum mechanically – either position or momentum.) However the similarities are more striking than the differences. The great thing about differential equations is that they produce nice smoothly continuous change in the quantities they describe. There is, therefore, none of the fitful air of discontinuity about the Schrödinger equation which we have come to associate with the quantum world. How then does probability come to rear its ugly head if our dynamics is based on a nice and deterministic equation like [9]? The answer is that the Schrödinger equation is only half the dynamical story.

If a quantum mechanical system is left to go its own sweet way without interference equation [9] will take care of its development. All will be smooth and determined. However, the act of measurement, of actually observing the system from the outside, involves a traumatic intervention. It is this act which introduces the probabilistic element, the jarring discontinuity in the system's experience. We spoke of Dirac's piece of chalk, or more realistically of an electron, in a state which is a superposition of the possibilities 'here' and 'there'. It is only when we put to it the rude experimental question 'where are you?' that it is forced to make the sharp choice between those two possibilities. Up to that moment it can be in a state evolving smoothly according to the Schrödinger equation, gently and continuously trimming the balance between 'here' and 'there'. At that moment of experimental interrogation it must choose the stark alternative of one or the other. (I hope you do not find this anthropomorphic language offensive. It is not to be taken seriously – of course, electrons do not 'choose' – but it is somehow irresistible for physicists to talk in this way, at least informally when they are not writing grave papers for their colleagues. Perhaps it is simply a need to humanise what is, after all, rather an inhuman subject.)

A measurement involves the registration of the result by some macroscopic device operating in the world of everyday experience. In conventional language we talk of a pointer moving across a scale to come to rest at the mark saying 'here' or at the mark saying 'there'. Otherwise we, who after all live in the everyday world and not at the atomic level, would be unaware of the observation. Pointers are reliable objects for the description of whose behaviour dependable classical physics is surely adequate. They do not fluctuate all over the place in an uncertain fashion. If, therefore, the pointer reads 'here' and we blink our eyes and look at it again we shall certainly find the pointer the second time still stolidly sitting at the mark 'here'. Now logically that second look is a second observation. The first observation might have given the answers 'here' or 'there'; sometimes it gives one and sometimes the other. But once that first observation has settled for 'here' on a particular occasion then there is no question that the second observation will agree with it and say 'here' again. Do you see that something remarkable has happened? Originally the electron was in a state of uncertain position; it might be 'here' or it might be 'there', with certain probabilities. However once its position has been determined it is in a totally different state, one of definite position. In the language of p. 27 the electron originally was not in an eigenstate of position but in some superposition of such states; after the act of measurement it finds itself in an eigenstate of position corresponding to the eigenvalue which is the result of that particular measurement.

We encountered a similar example of this happening when we discussed the transmission of photons by a crystal of tourmaline (p. 19). The crystal acts as an analyser which ascertains whether or not the photon has polarisation perpendicular to the optical axis of the crystal. The photons which approached the crystal were in a state which was a superposition of that state of polarisation together with the state of polarisation parallel to the optical axis. The transmitted beam consists only of photons with the perpendicular polarisation. Being let through the crystal is tantamount to a polarisation measurement and it changes the photon's polarisation state in this discontinuous way (from x' to y, in our earlier notation).

Every act of measurement has this character of entailing instant change. Beforehand our system is not in general in an eigenstate of the observable we are intending to measure but rather it is a superposition of such states. Afterwards the system is in that particular eigenstate, selected from the original superposition, which corresponds to the eigenvalue actually obtained as the result of that measurement. In the jargon of the subject this discontinuous change is called the *collapse of the wavepacket*. [The idea is that probability, which was originally spread out in a wavefunction (or packet) covering 'here', 'there' and perhaps 'everywhere', is now all concentrated 'here'. It has collapsed in on itself.]

The dynamical development of a quantum system is, therefore, made up of two elements. In between measurements its state vector (wavefunction) propagates in the smooth and orderly way prescribed by the Schrödinger equation. In this respect things are not qualitatively different from the picture afforded by classical physics. At least there is continuity of development. However when a measurement is made the state vector undergoes a discontinuous change, collapsing onto that particular eigenvector which corresponds to the result actually observed on that occasion. At this point is located all that is capricious and probabilistic, all that is quintessentially quantum mechanical and alien to classical mechanics. Here the smooth train of causality is interrupted, the statistical rot sets in. The seeds of this development were sown when we allowed the strange possibility of superposing states, mixtures of the classically immiscible 'here' and 'there'. Those seeds came to full flower when we attempted to analyse the consequences of the act of measurement.

It will be clear that these two dynamical elements, the smooth propagation à la Schrödinger and the discontinuous collapse on making a measurement, sit uneasily side by side. As prescriptions they are crystal clear. In performing calculations we know exactly what to do and the answers fit nature like a glove. However when we attempt to unpack what is involved in measurement and to see how its discontinuity is consistent with the Schrödinger equation, then the puzzles and disputes to which quantum mechanics is heir become apparent. It is the task of the chapters that lie ahead to go into these matters. In the course of that discussion we shall also try to understand the picture given us by quantum theory of the nature of physical reality.

On this last point let me for the present be content with a simple observation. Your average quantum mechanic is about as philosophically minded as your average garage mechanic. However, just as the latter might be thought to have an intuitive grasp of affairs automotive which ought to be taken into account by anyone prone to theorising about the motor car, so it is conceivable that the way theoretical physicists regard the objects of their study might be a factor to be taken into account in assessing their significance. I think that it is true to say that almost all practitioners of quantum mechanics talk in a strongly realistic way about a world described by wavefunctions. They do not speak about electrons and their states of motion as if they were to be taken less seriously than, say, billiard balls or elephants. Of course, critical analysis might show this to be due to naive misapprehension. But I submit that it might be wise to look for an interpretation of quantum mechanics which comes as near as possible to being in accord with the attitude so widespread among its users.

Chapter 4

Which way did it go?

One of the most remarkable members of the generation of physicists who came into prominence in the years following 1945 was Richard Feynman. He invented a diagrammatic technique for marrying quantum mechanics to special relativity which has proved a powerful source of insight and employment for many of us. Ebullient in character, a Nobel laureate retaining more than a touch of the New York kid, Feynman is keenly interested in teaching physics as well as creating it. Some years ago he gave a sequence of exciting and idiosyncratic lectures on physics to students at Caltech. They were subsequently published as *The Feynman Lectures in Physics*. The third and final volume is devoted to quantum theory and there he tells us that its first chapter will tackle

the basic element of the mysterious behaviour in its most strange form. We choose to examine a phenomenon which is impossible, *absolutely* impossible, to explain in any classical way, and which has in it the heart of quantum mechanics. In reality it contains the *only* mystery. We cannot make the mystery go away by 'explaining' how it works. We will just *tell* you how it works. In telling you how it works we will have told you about the basic peculiarities of all quantum mechanics.

After such a whetting of the appetite, let us also consider what happens when you fire objects at a screen with two slits.

Schematically the arrangement is as shown. It involves three elements: a source of the objects, the screen with its slits and a second screen which incorporates some means of detecting the objects which hit the different parts of it. Now consider three contrasting experiments with such a piece of apparatus.

In the first experiment our source is a not very accurate gun which sprays bullets onto the first screen. The bullets may bounce off the sides of the slits or go plumb through the middle, so that we cannot be certain of exactly where they will hit the second screen. There will be some spread in the possible point of impact. By patiently

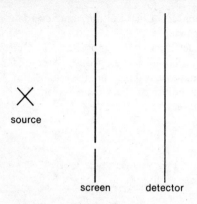

source

screen   detector

watching and recording hits we can build up a probability profile which will indicate how likely it is for a bullet to hit any particular point on the detector. (If a very large number of bullets are fired the probability profile has exactly the same shape as the profile recording the number of hits.) If the width of the slits and the spread of the bullets are appropriately chosen we may get a profile like this:

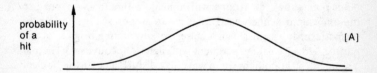

probability
of a
hit   [A]

with a hump at a point on the detector equidistant from the two slits. It is made up by adding together the profiles corresponding to bullets which have gone through slit 1 and bullets which have gone through slit 2:

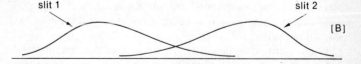

slit 1   slit 2
                                                                  [B]

We see that the humps in the profiles B are displaced towards the corresponding slits and that the single symmetrical hump in A is built up by adding together these two displaced humps.

Of course, our bullets arrive at the detector one by one and each of them has traversed one or other slit. All this is crashingly obvious. It corresponds to our classical intuition, and everday experience, of how particles behave.

For our second experiment let us take the opposite tack. Our source will now produce waves. Perhaps it is an acoustic source of sound waves. Our detector in that case will measure the loudness of the signal received at different points of the second screen. The resulting loudness profile will look like this:

[C]

Again there is a hump at a point equidistant from the two slits but the pattern oscillating away from it is totally different from the particle case A. This is due to the characteristic wave property of *interference*. At some points on the detector a crest from slit 1 will coincide with a crest from slit 2. In that case the two waves reinforce each other and one gets a large effect, a hump in the pattern. At other points the crest from 1 will coincide with a trough from 2 so that they almost cancel each other out to give one of the hollows of the pattern. It was just such a pattern of alternating brightness and darkness in a two-slit experiment with light that convinced Thomas Young in 1803 that light was a wave-like disturbance. It will be clear that the presence of *two* slits is esential to give an interference pattern. If I close one slit the pattern disappears since there are no longer two sets of crests and troughs to reinforce or cancel each other.

In this experiment the waves beat continuously all along the shore of the detector. There is none of the discrete lumpiness we saw in our first experiment when a bullet at a time hit the screen, now here, now there. The sound waves come *legato*, not *staccato*. Once again, everything we have said is crashingly obvious to the well-versed classical physicist.

For our third experiment we must move into the quantum world. Our source will spray the slitted screen with electrons and we shall have some means of detecting them – a Geiger counter array maybe – at the second screen. Wise as we now are in the ways of quantum theory we can, perhaps, anticipate the mystery of which Feynman spoke. Will it not turn out to reside in a strange mixture of particle-like and wave-like behaviour?

It will indeed. As we listen to the Geiger counters we shall hear a succession of clicks. The electrons are arriving at the screen discretely, one by one, just like the bullets of our first experiment. This is the particle character asserting itself. However when we accumulate enough data to be able to plot a probability profile it will not look like the smooth single-hump curve of A but instead it will resemble the oscillating interference pattern of C. This is the wavelike side of the coin.

The sixty-four thousand dollar question is this: as the electrons arrive one by one at the detector, which slit have they come through? Click! Suppose that a particular one came through slit 1. Then for it slit 2 was irrelevant; we could momentarily have closed it up. However, if we actually perform a single-slit experiment with 2 closed we do not get an interference pattern at all. Instead we get the same sort of smooth hump which constitutes slit 1's contribution to B. The result is scarcely surprising. We have seen how the phenomenon of interference depends on the collaborative reinforcement or cancellation of two trains of waves. With a single slit there would only be one such train. Therefore the pattern C cannot be due to an electron which traverses slit 1, nor, of course, to an electron which traverses slit 2.

There is only one way out of the dilemma. As Sherlock Holmes said, 'When you have eliminated the impossible whatever remains, however improbable, must be the truth.' The indivisible electron came through both slits! In the language we have learnt to talk, its state of motion is a superposition of the state in which it passes through slit 1 and the state in which it passes through slit 2. What is classically inconceivable is quantum mechanically inescapable.

Such is our conclusion if we passively let things take their course. Suppose, however, we take active steps to determine which slit the electron traversed. Feynman discusses the idea of putting a lamp between the two slits to illuminate the electrons as they pass by. Just before each Geiger counter click we see a flash of light from the illuminated electron. Sometimes that flash is near slit 1 and in that case we know that the electron actually went through the first slit. Sometimes the flash is near slit 2 with the conclusion in that case that the electron went through the second slit. Thus we are able to pin things down. Great! But by doing this we have changed the experiment and we had better check whether the result has also altered. In fact it has. Our intervention with the lamp has destroyed the interference pattern. Instead of C we now get A again. The measurement with the lamp has enforced, as every measurement must do, a collapse of the wave packet. The flash near slit 1 meant that our electron was no longer in a superposition of states but was uniquely in the state corresponding to traversing slit 1. Without superposition there is no interference; C becomes A, the sum of the two terms B, one for each slit.

The two-slit experiment neatly underlines a basic feature of quantum theory, that we only know where things are (which slit they go through) if we actually look and see. Otherwise they are not locatable. Classical physics is played out before an all-seeing eye. It has trajectories. We can follow the motion of particles from moment to moment as they interact with each other. We have to give that up in the quantum world. There are no longer trajectories. This has important consequences for the discussion of identical particles.

Suppose I have a brace of electrons. Each is a specimen of one of nature's standard products. They are identical in the same way that the articles coming off a well-engineered assembly line are identical. In classical physics identical particles are nevertheless distinguishable. By that I mean that, though they do not carry intrinsic labels to discriminate them, yet they can be given extrinsic labels for that purpose. I watch carefully the articles coming off the assembly line and remember that that one was the first, that the second and so on. A classical physicist can decide at the beginning of an experiment to call that electron over there A and this one over here B. Subsequently they may interact and go through all sorts of complicated convolutions, but because he can in principle follow the motion the classical physicist's labels have an abiding significance. Tracking along the trajectories he can establish that the electron which eventually fell onto this spot of the photographic plate was the one he called A whilst the one that collided with the gold atom was the one he had called B. Quantum mechanically (that is to say, with the electrons of the real world) this cannot be done. The electrons are not only identical, they are also indistinguishable. At any one instant I can call the electron there A and the electron here B if I want to. However the nomenclature is fleeting. Since there are no trajectories to follow I cannot tell when I next see an electron whether it is the one I called A or the one I called B or some other one altogether. I cannot, therefore, say that subsequently A hit the photographic plate and B collided with the gold atom, but simply that *an* electron hit the photographic plate and *an* electron collided with the gold atom. We have been talking about electrons but obviously exactly similar considerations of indistinguishability apply to all kinds of identical particles in the quantum world.

Indistinguishability is bound to produce consequences for the structure of the theory. What those consequences are is not too hard to deduce (see Appendix, A7) but, ever conscious of the susceptibilities of my mathematically weaker brethren, I shall content myself with a qualitative discussion. It turns out that there are two possibilities, some sorts of identical particle choosing one alternative and some sorts the other. Suppose I have a particle 'here' and an identical particle 'there'. Because I cannot distinguish one from the other it must be a physically identical statement to say that I have a particle 'there' and an identical particle 'here'. Perhaps you can see that this must mean that interchanging the two particles must leave me in exactly the same physical state. *Plus ca change, plus c'est la même chose.* I must now confess something which I kept back from you in Chapter 3. The correspondence between physical states and wavefunctions (state vectors) is not strictly unique; it is not one-to-one as the mathematicians say. The rules for calculating physical consequences from the mathematical formalism are such that several wavefunctions, provided they differ from each other only by *trivial* multiplicative factors, give the same results. In particular multiplying the wavefunction by -1 does not change the physics at all. The wavefunctions ψ and $-\psi$ describe the same state of motion. This means that it is not necessary to require that the *wavefunction* is unchanged by the interchange of two identical particles but simply to specify that if it does change it does so in one of these trivial, physically equivalent, ways. It turns out that there are two allowed possibilities. Either identical particles are *fermions*, which means that their wavefunction has to change sign under interchange, or they are *bosons*, in which case they have the even simpler property that their wavefunction is completely unaltered by an interchange. In the jargon of quantum mechanics, this choice of behaviour is called the *statistics* obeyed by the particle. All particles of the same species obey the same statistics.

For example, all electrons are fermions. They are said to obey Fermi–Dirac statistics. From this it follows that they satisfy the *exclusion principle*. Discovered by Wolfgang Pauli, this asserts that no two electrons are ever in exactly the same state of motion. The presence of one electron in a specific state (for example, at a particular point in space) excludes all others from it. Electrons are like spouses in monogamous countries; there is either one or none. The argument deducing this from Fermi statistics is rather straightforward. Suppose two electrons *were* in the same state, let us say both 'here'. Interchanging them would mean that instead of having an electron 'here' and an electron 'here' I had an electron 'here' and an electron 'here'. In other words no change has taken place at all. But Fermi requires there to have been a change of sign. There is only one quantity clever enough to be both itself and minus itself. That quantity is zero. But zero means the wavefunction vanishes and there is no state at all. In other words we were mistaken in thinking that there could be two electrons 'here', or in any other state of motion for that matter.

Photons, on the other hand, are bosons. They obey Bose–Einstein statistics. Now there is no bar to having more than one particle in each state. Quite the contrary. It turns out that bosons positively like to be in the same state together. They are like money; the more you have the easier it is to get extra. (This property is the basis of the maser and the laser.)

Although the interference pattern C and the single hump patterns of B look about as dissimilar from each other as one could imagine, there is in fact a hidden mathematical connection between them. This is worth unravelling, even at the expense of a brief excursion into the mathematical realm of complex numbers.

A typical complex number z is a combination of an ordinary number with some multiple of i, the square root of minus one. That is to say, we can write it in the form

$$z = x + iy, \tag{1}$$

where x and y are real (that is, 'ordinary' numbers). An important quantity associated by the mathematicians with the complex number z is its modulus, written $|z|$ and defined as

$$|z| = +\sqrt{(x^2 + y^2)}. \tag{2}$$

I have explicitly indicated that the positive value is to be taken for the square root.

In quantum theory probabilities are always calculated in a two-step fashion. First one calculates a complex number a, called a *probability amplitude* (see Appendix A5) and then the probability itself (which, naturally, has to be a positive number and less than 1) is given by the square of the modulus of a, or $|a|^2$ as we have learnt to write it. The wavefunction $\psi(x)$ in Schrödinger's wave mechanics is a particular example of a probability amplitude, giving in that case the probability of finding a particle at a specified location x.

The two probability distributions of B are given by the squares of moduli of two such wavefunctions. If ψ_1 is the wavefunction at a point on the detecting screen corresponding to the electron's having traversed the first slit, and P_1 the associated probability, and if ψ_2 and P_2 are the similar quantities for slit 2, the rules I have just stated imply

$$P_1 = \mid \psi_1 \mid^2 \text{ and } P_2 = \mid \psi_2 \mid^2. \qquad [3]$$

Then, in the case when both slits are open, the probability (P_{12} let us call it) is just given by

$$P_{12} = \mid \psi_1 + \psi_2 \mid^2. \qquad [4]$$

The quantities ψ_1 and ψ_2 are what we called probability amplitudes (p. 88). The squares of their moduli give the actual probabilities, as equation [3] says. In equation [4] we see the superposition principle again at work, adding (as it should) not probabilities, but probability amplitudes. The oscillations of C are due to the alternating reinforcement and cancellation between ψ_1 and ψ_2. I think you can see that [4] is the quantitative version of the statement that to get C the electron must go through both slits. Adding wavefunctions is superposing states.

I have just given you the traditional quantum mechanical recipe for calculating probabilities. It assumes that the wavefunctions ψ_1 and ψ_2 were obtained by solving the Schrödinger equation with slits 1 and 2 respectively open. There is, however, an alternative way of proceeding. Richard Feynman invented a method for formulating quantum theory, identical to the traditional approach in its physical consequences, which bypasses the Schrödinger equation and proceeds directly to the calculation of probability amplitudes. It is called the *path integral* or *sum over histories* approach.

To calculate ψ_1 Feynman tells us that we should think of all the different ways in which an old-fashioned electron with classically picturable simultaneous position and momentum could travel from the source through slit 1 and onto the specified point on the second screen. There is obviously a vast number of such possible trajectories. Some are direct and made up of straight lines, others meander around; in some the electron moves fast, in others it dawdles. Each is called a path or a history. (Both terms are used.)

Feynman tells us to consider all such possibilities and to assign to each a complex probability amplitude. He is able to specify a rule for what that amplitude should be. It involves a quantity which physicists call *action*. We need not concern ourselves with its detailed definition, though it is precise enough for those in the know. It would not be too misleading to say that it lives up to its name and represents a measure of the degree of 'business' of the electron following that particular path. A suggestive thing about it is that action can be measured in units of Planck's constant $\hbar$. It has the same 'dimensions' as the physicists say. (You can measure the length of a cricket pitch in terms of the height of a tankard, if you wish, because they are both lengths, but you cannot specify it in terms of pints because volume and length are of different dimensions.) Thus there is a natural number associated with any quantity of action, namely the number of $\hbar$ units – $A/\hbar$ that is – which does not have to be a whole number and usually isn't. Feynman next gives a rule for associating a complex amplitude with this number. [For the learned it is $\exp(iA/\hbar)$.] You then add together all the contributions from all the different paths and – hey presto! – the result is the same probability amplitude which you would have calculated by the more pedestrian procedure of solving the Schrödinger equation.

That may sound rather complicated. In fact it *is* complicated, immensely so, because the sum over paths or histories is a gigantic and rather ill-defined notion. Its proper specification has only been given for some simple cases. For calculational purposes the Feynman approach is an unwieldy steam-hammer only capable of cracking some particularly brittle nuts. Such power as it possesses is qualitative rather than quantitative. It provides an alternative and bizarre answer to the question 'Which way did it go?', namely 'Every which way'. We can think of the electron not only as going through both slits but also as following paths both direct and indirect, moving both rapidly and slowly. From the point of view of conventional quantum theory the electron has no trajectory; from Feynman's point of view it has every trajectory. Either way the neat classical idea of tracing a well-defined motion is lost.

Feynman's approach gives a particularly simple way of understanding how that neat classical picture will nevertheless emerge as a limiting case for 'large' systems, that is for systems whose action is very big on the scale set by $\hbar$. It is, of course, essential that this happens. No really successful physical theory ever completely disappears from the scene. The worst that can happen is that its domain of applicability becomes circumscribed. When we want to calculate the orbits of Voyager satellites in the space programme Newtonian mechanics is more than accurate for the purpose. It is essential for quantum theory that it should be able to annex to itself the triumphs of its predecessor. It is no good explaining the microscopic at the expense of the macroscopic. Quantum theory must produce the same goods in the large-scale domain as those of Newtonian mechanics which works so well. The pioneers of the subject were very conscious of this constraint on their endeavours. One of their guidelines was what they called the *correspondence principle*, the requirement that classical mechanics should be recoverable for large systems.

Feynman offers us a simple way to see that this happens. In his gigantic sum over histories there is a vast amount of interference between the contributions of neighbouring paths which have a tendency to cancel each other out. The larger the size of A in units of h the more rapid will be the variation and the more sweeping the resulting cancellations. For really large systems this will have the consequence that the only paths which contribute significantly to the final result will be those in a region where the action changes as slowly as possible, since here the cancellations are minimised. This region is threaded along what is called the path of stationary action, since the latter is by definition the path from which small changes of trajectory produce negligible changes of action. But in 1744 Maupertuis had shown that classical trajectories are just these paths of stationary action. (He was motivated by a belief that an 'economical' God would wish to organise things in this way. We may think that he was over-confident about his degree of insight into the divine nature, but it was a capital discovery all the same.) Anyway Feynman makes it clear that classical mechanics will arise as the limiting behaviour of large systems since the only paths that will count in their sum over histories will be extremely close to the classical trajectory of least action.

Sums over histories also played an important psychological role in the development of physics since they led Feynman to propose his diagrammatic ideas about which I spoke so warmly at the start of this chapter. However they were not indispensable for that purpose. Schwinger and Tomonaga – who shared the Nobel prize with Feynman – produced equivalent, if less picturesque, versions of the same formalism by more conventional arguments. Nowadays no one initiating a novice into the pleasures of Feynman diagrams would chose to do so from the sum over histories point of view.

Chapter 5

More-or-less knowledge

The one thing that everyone knows about quantum theory is that it contains Heisenberg's uncertainty principle acting as a limit upon our powers of exact knowledge. Instead of the clarity and precision of Newtonian mechanics, we have to be content with a more fuzzy account of affairs. We found some hint of this in Chapter 3 when we noted (p. 28) that not all observable quantities could be measured simultaneously. That conclusion arose from a mathematical fact (that operators did not commute), via rather abstract arguments involving the idea that the eigenstates of hermitean operators are the state vectors corresponding to the precise measurement of the observables the operators represent. This was the way in which uncertainty asserted itself in Heisenberg's original formulation of quantum mechanics. To make assurance doubly sure – and also to begin a move in the direction of greater picturability – let us now see how Schrödinger would have obtained the same result.

His approach is based on the wavefunction ψ, which gives (via $|\psi|^2$) the probability of finding a particle at a particular point. Since we shall have to be content with inexact knowledge we will not insist that our electron is to be found at a definite point but simply require that it is localised within an interval of space of length Δx. This means that we shall have to choose a ψ which is only non-zero within a range Δx. Such a wavepacket (as it is called) is illustrated in the figure. It cannot be made up of a wave with a unique value for its wavelength since such a wave stretches on for ever. Instead it has to

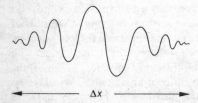

Δx

be made up of a band of waves of different wavelengths, cunningly chosen to cancel each other out outside the region of width Δx and to reinforce each other inside it. It may strike you as rather remarkable that this is possible, but the mathematicians assure us that it can be done. Their study of such questions is the subject of Fourier analysis. I think it will be clear that the narrower the space Δx, the more stringent the cancellation requirement outside it, and in consequence a greater number of different waves will have to be added together to achieve it. In other words, the width of the band of wavelengths thus required will be inversely proportional to Δx; as the latter narrows, the former must widen.

Now, de Broglie, when he hypothesised matter waves, related their wavelength to the particle momentum through the equation

$$p = h/\lambda. \tag{1}$$

This interpretation was taken over by Schrödinger when he invented wave mechanics. Thus in particle terms, a broader band of wavelengths means a wider range of momentum values. In other words, as the location is better specified (Δx decreasing), the momentum is further spread out (the uncertainty Δp increases), and vice versa. All this can be put on a quantitative basis and doing so results in the celebrated condition written

$$\Delta x \cdot \Delta p \gtrsim \hbar; \tag{2}$$

that is, the product of the uncertainties in position and momentum is always at least of the order of magnitude of Planck's constant. In the extreme case of x being exactly known ($\Delta x = 0$), the momentum must be totally unknown (Δp infinite), and vice versa.

At first encounter this conclusion was exceedingly unpalatable to those brought up in the clarity of classical physics. It meant leaving firm ground for a quivering quantum quagmire. The old clear-cut certainties dissolved into an indeterminate haze. The man with the courage to face up to the issue was Heisenberg.

The way that so far we have approached the matter has been highly theoretical. Our talk has been of non-commuting observables or the Fourier analysis of wavepackets. Mathematics is a wonderful subject but the physicist has always to ask himself whether he is using those mathematical constructs which are truly appropriate to the way the world is. *Mathematically* we cannot evade equation [2] as a consequence of the formalism we have adopted. It still remains a question whether it all works *physically*.

The answer to that vital second question is only partly empirical. If Heisenberg's matrix mechanics, or Schrödinger's wave mechanics, had not given the right answers for the behaviour of objects like hydrogen atoms then certainly no one would have got very excited about them. Nevertheless, there is also the question of whether there is an inner physical consistency in the interpretation proposed which matches the undoubted mathematical consistency of the equations which express the formalism. This was the problem to which Heisenberg addressed himself. The uncertainty principle [2] alleges a certain limit on the accuracy with which position and momentum can simultaneously be known. Heisenberg realised that the formalism also implicitly prescribed what can be measured and how well. Did these two prescriptions tally?

Faced with such a question the theorist has to repair to a mental laboratory where he conducts *thought experiments*. By that I mean that he considers schematic ways of making measurements, set up in accord with the rules of the theory, and sees if by any ingenious means he can circumvent a restriction like that imposed by equation [2]. From such a course of imaginative practicals Heisenberg emerged with the triumphant conclusion that his uncertainty principle was a consistent epistemological consequence of quantum theory. His method can be illustrated by an account of the most celebrated experiment in his mental laboratory manual – the *γ-ray microscope*.

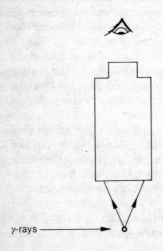

γ-rays

We are endeavouring to measure as accurately as possible the position and momentum of an electron. As far as the former is concerned we decide to find out where the electron is by shining a beam of light onto it and taking a look through a microscope. If none of the light bounced off the electron into the microscope there would be nothing to see. There must, therefore, be some interaction between the radiation and the electron, and the latter's state of motion is perturbed in consequence. If we want as accurate knowledge as we can get we must minimise the degree of disturbance caused by the intervention of the act of measurement. In classical physics there is, in principle at least, no limit to how gentle we can make the perturbation. This is because classical beams of radiation can be of arbitrary weakness. That may cause practical problems in detecting the consequences, but this is a thought experiment after all, and we can assume that technical ingenuity will prove equal to the task of providing adequate amplification of the signal. There is, however, one classical restriction which we must take into account, namely the resolving power of optical instruments. No microscope can produce an absolutely sharp point image. Rather it is necessarily fuzzed out by an amount which is of the size of the wavelength of the light employed. The answer to that is simply to make sure that we use radiation of a wavelength short enough to be compatible with whatever accuracy we choose to specify. That is why the instrument is called the γ-ray microscope; γ-rays are short-wavelength rays of light. They are beyond the visible spectrum but we do not have to make our observations with the naked eye and there are ways in which γ-rays can be detected to form an image pattern. Our classical conclusion is, therefore, the familiar and expected one. We acknowledge that measurement imparts some disturbance to the system observed but we see no barrier of principle to reducing this to any arbitrary assigned degree of smallness. Exact knowledge is within our grasp, as near as makes no difference.

Quantum mechanically the situation is completely different owing to the existence of the photon. There is now an irreducible minimum for the strength of the beam, namely that it contains at least one photon! According to Planck's condition (p. 6) a single photon of frequency v carries energy hv. Because the microscope gives an image which is fuzzy over a distance of the size of the wavelength of the light being used, if we want to measure the position more accurately we must decrease the wavelength. But the wavelength multiplied by the frequency is always equal to the velocity of light, so this decrease in wavelength has the result at the same time of increasing the frequency ν. Inevitably we thereby increase the energy carried by a single photon and make its interaction with the electron correspondingly more rumbustious (with the effect of increasing the degree of uncontrollable disturbance to its momentum). A quantitative evaluation of the effect leads straight to the uncertainty relation of equation [2] (see Appendix, A8).

Without going here into the details of the calculation there are some interesting points which can be made about it. The mere fact that the photon's colliding with the electron and bouncing off it into the microscope has the effect of changing the electron's momentum would not of itself be disastrous. If we could calculate accurately the consequential change in the electron's momentum then we could allow for the effects of the collision and our knowledge would still be capable of being made exact. It is not disturbance as such but *uncontrollable* disturbance which leads to uncertainty. The reason that we cannot calculate the necessary correction is that we do not know which way the photon went when it bounced into the microscope. All we can say for sure is that it entered the microscope, perhaps on one side, perhaps on the other, perhaps in the middle. Aha! you say, the solution is clear. Narrow the aperture of the microscope and then we shall have as accurate a knowledge of the photon's direction as we please and thus the uncertainty in the momentum transferred to the electron will be controlled. That is certainly possible but a penalty is exacted for it. The resolving power of a microscope depends not only on the wavelength of the light but also inversely on the aperture of the instrument. To be sure working with a narrower aperture will decrease Δp but it will bring about a compensating uncertainty Δx in the electron's position. This proves to be in perfect accord with the requirements of equation [2].

Heisenberg did not immediately appreciate this last point and in the first draft of his paper it was not considered. Niels Bohr pointed it out to him and he had to add a note in proof to deal with the question. Behind that incident there lies an amusing story. A little earlier Heisenberg had taken his Ph.D. at Munich. The professor of experimental physics, a grand old man called Wien, was displeased that this brilliant young theorist had adopted a rather cavalier attitude to the laboratory classes which were provided. He decided, therefore, to submit the young whipper-snapper to a tough oral examination with searching questions of an instrumental nature. The first questions Wien asked were related to the resolving power of optical instruments. Heisenberg was totally unable to answer them. Wien decided that the candidate should fail. Artur Sommerfeld, the professor of theoretical physics, was aghast at this fate for a young man whose brilliance was already apparent. After tricky negotiations between the two professors a judicious compromise was reached by which Heisenberg was allowed to pass but at the lowest level. The young turk of theoretical physics could not have guessed that this very point in classical optics, probed by Wien, was to be of importance in one of the most fundamental papers he was subsequently to write. In later life Heisenberg was to admit ruefully that Wien was right to press his point, though we may be sure that Sommerfeld was right to press his point also.

The γ-ray microscope shows how central particle – wave duality is to quantum mechanics. When we calculate the uncertainty in position, Δx, we use formulae for resolving power which are derived from wave optics. When we calculate the uncertainty, Δp, in the momentum transferred to the electron by the photon, we treat their collision in a particle way, just like a collision between two billiard balls. The experiment also illustrates how thoroughgoing one has to be in applying quantum mechanics. One might be tempted to argue that the photon is absorbed by the microscope and that by measuring the latter's consequent recoil, due to acquiring the photon's momentum after its collision with the electron, we could evaluate this momentum as accurately as we wished. (Remember these are 'in-principle' thought experiments, not practical procedures, that we are discussing!) If the photon's momentum were thus determined the electron's uncertainty Δp could be made as small as we pleased, thus beating, it would seem, the restriction [2]. In fact this does not work because we have to treat the microscope as also subject to quantum mechanical uncertainty. If we were to measure its recoil momentum sufficiently accurately to limit the electron's Δp significantly, then Heisenberg would require the *microscope's* position to be compensatingly uncertain. This would blur the image and increase the electron's Δx, again in perfect accord with equation [2]. We cannot nibble at quantum theory. If we are to digest it properly it must be swallowed whole. This will prove an important point to bear in mind in the discussions of Chapter 6.

A similar attempt to cut the Gordian knot of Chapter 4 also proves to be fallacious. We might suppose that we could pin down which slit the electron goes through by mounting the slits independently so that the one that deflects the electron onto the screen will signal this by recoiling. However the accurate knowledge that we would need of the slit's momentum will so blur its position that the diffraction pattern C, which depends on where the slits are, would be washed out. The conclusion is, therefore, the same as that found with Feynman's lamp device. If you do succeed in fixing which slit the electron went through you pay the penalty of having lost the interference effects. These tests of physical consistency give one great confidence in the interpretative scheme proposed for the formalism.

So far we have just talked about position and momentum. They are simply particular examples of what students of dynamics (classical or quantum) would call a co-ordinate q and its conjugate momentum p. Two such quantities q and p are always related physically by the fact that their combined dimensions are those of action; that is, the product $q \cdot p$ can always be assigned a value in units of Planck's constant $\hbar$. It turns out that Heisenberg is universally limiting and requires that the uncertainties of any such pair of quantities should satisfy

$$\Delta q \cdot \Delta p \gtrsim \hbar, \tag{3}$$

in obvious generalisation of [2].

A particularly interesting pair of variables to which to apply this is provided by the energy E and the time t:

$$\Delta t \cdot \Delta E \gtrsim \hbar. \tag{4}$$

At first sight this is a puzzling relation, because time is not quite an observable property of a system in the normal sense but is rather a parameter marking when an observation is made. The great Russian theoretical physicist Lev Davidovitch Landau was fond of saying about [4] 'There is obviously no such limitation – I can measure the energy and look at my watch; then I know both energy and time!' The correct way to think about [4] is to interpret it in terms of energy transfers. A limitation is placed upon the accuracy with which one can specify the amount of energy transferred together with a knowledge of the time at which the transfer took place.

The time – energy uncertainty relation can be used to give a 'folksy' account of a characteristically quantum mechanical phenomenon – *tunnelling*. Suppose you shoot a marble and try to get it across a hump in the ground. As the marble attempts to surmount the obstacle it looses speed. In physics terms it is using up some of its energy of motion (its kinetic energy, as we say) in doing work against gravity in climbing the hump. (We call that acquiring gravitational potential energy.) If the marble is not moving fast enough it will run out of kinetic energy before it reaches the top, come to a momentary halt, and then roll back down again the way it came. However, if the marble has enough energy to take it to the top of the hump then it will, of course, roll down the other side. Classically one or other of these things happens to all the marbles that start out at the same speed; either they all get over or they all fall back. In the first case they had enough kinetic energy to make it; in the second case they did not. Quantum mechanically the same sort of experiment, using electrons instead of marbles and an electric rather than gravitational barrier, does not produce such clear-cut results. Just occasionally an electron whose energy is not enough to take it legitimately over the top will nevertheless emerge the other side. It has tunnelled through. How do these rare moles manage it? Is energy not conserved in quantum mechanics? The answer is, yes it is, but only up to a point. The relation [4] enables an electron to do a bit of creative dealing, borrowing energy uncertainty against time. If it is quick about it this will give the electron the chance to get over the hump and repay the loan as it emerges on the other side.

If you are a person of fastidious intellectual taste you will not much care for the comic-strip account given in the preceding paragraph. I certainly agree that I would not be wholly confident of the conclusion about tunnelling if I did not know that rigorous calculations with the Schrödinger equation lead to the same conclusion. However one should not despise too hastily such hand-waving discussions. They afford a modest degree of intuitive insight, which is always welcome. Doing theoretical physics is usually a two-step process. In the first place one tries to get some feel for what is going on. Only then can one turn with any success to the second task of turning that view into the formal quantitative language of calculation. The latter process is the exercise of a technique; the former is the exercise of imagination. Although the calculations can often be tricky and demanding – perhaps beyond one's power other than in some crude approximation – it is the first creative stage which is the difficult one. Any picturesque mode of thought which enhances intuition is to be grasped gladly.

Tunnelling played an important part in the early history of quantum mechanics. George Gamow used the idea in 1928 to explain the α-decays of heavy nuclei. Such nuclei behave as if they have α-particles rattling around inside them. These particles occasionally shoot out, having penetrated an energy barrier which would classically be insurmountable. The puzzle was that the times one had to wait for this to happen varied very greatly from nucleus to nucleus, despite there being only small changes in the circumstances involved. For example Radium C′ obliged after only a thousandth of a second whilst the not so different Thorium took, on average, ten thousand million years. Perhaps our account of the wizardry of energy juggling involved in the tunnelling process makes it not altogether implausible that a doubling of the height to be scaled could immensely increase the difficulty of the transaction. Anyway, that is the case and Gamow showed that the quantitative formalism accounted for the experimental results in a highly satisfactory manner. It was an achievement of greater significance than merely explaining some puzzling data. It indicated for the first time that quantum theory, which had been developed to account for atomic phenomena, still remained successful when applied to nuclei, systems which are a hundred thousand times smaller.

Throughout those heady years in the middle twenties in which the young men were creating modern quantum theory, Niels Bohr played a role which was a mixture of elder statesman and father-confessor. At his institute in Copenhagen ardent youths would arrive to lay their latest intellectual offering before him and to seek his criticism and approbation. Countless reminiscences recall the uniquely stimulating and searching atmosphere at the court of this philosopher-king. There is no doubt that Bohr's influence was immense, but his contribution at this period was largely implicit rather than explicit, as far as the tale of attributed discoveries goes. There is, however, one notion which is the visible tip of Bohr's intellectual iceberg, an idea explicitly associated with his name. It is called the *principle of complementarity*. In essence it is very simple. Heisenberg's relation [2] enforces upon us a choice if we wish to attempt a description of physics in terms of exact quantities. Either we can know the position exactly, in which case we sacrifice all specification of momentum, or we make the converse choice. Either extreme is possible and the experimental arrangements necessary to realise these choices are as totally distinct as are the formalisms which express them. A position description of nature, or a momentum description, are available to us. We can have one or the other but not both simultaneously. In a word, they are complementary.

In later life Bohr tended to exalt this useful observation of alternative and mutually exclusive descriptions into a grand philosophical principle. It can cover a multitude of incompatibilities. Take the ever-puzzling problem of the relation of biology to physics. Living systems certainly have physical components but you have to be a very thoroughgoing reductionist to feel happy with the notion that biology (and anthropology) are nothing but, admittedly elaborate, corollaries to physics. In our perplexity Bohr is at hand to help us. He points out that if I tear a living system apart into its component atoms, I kill it. There is a complementarity between life and atomic physics.

Well yes, but in those sorts of terms complementarity becomes a passkey which turns suspiciously many locks. The point is that in its quantum mechanical form it deals with a mutual incompatibility which we understand rather well. We can study in detail how the complementary descriptions of position and momentum relate to each other. On broader questions, like physics-and-biology, our understanding is considerably less and complementarity is in danger of becoming a descriptive catchword rather than an interpretative principle. Biophysics is certainly able to cast significant light on processes occurring within living cells. The successes of molecular biology are sufficiently striking to show that whatever the true relation of biology may be to physics, it is certainly a subtle one that cannot be encapsulated in a single word.

A theory with an uncertainty principle is not going to yield a determinate dynamics in which position and momentum are the subjects of clear-cut prediction. This feature of quantum mechanics proved very distasteful to some of the very men who had helped to create the subject. In discussions with Bohr in September 1926 Schrödinger said 'If we are going to stick to this damn quantum-jumping, then I regret that I ever had anything to do with quantum theory'. Louis de Broglie also tried from time to time throughout his later life to find ways of reconciling quantum mechanics with a more deterministic picture. But the man who reacted most violently, and was never fully reconciled to this aspect of the theory, was one of its intellectual grandfathers, the great Albert Einstein, whose explanation of the photoelectric effect (p. 6) had been a key step in establishing the existence of the photon. In 1924 Einstein had said that if the ideas, then in the air, of renouncing strict causality proved to be correct he would 'rather be a cobbler, or even an employee in a gambling house, than a physicist'. Later, in a letter to Max Born, he delivered himself of his celebrated remark that he did not believe that God (whom he customarily referred to in comradely terms as 'the Old One') played at dice.

Einstein, therefore, set to work to try to demolish the accepted version of quantum mechanics. The point to which his assault was directed was the distasteful uncertainty principle. He sought to show by ingenious argument that it must be false because he could find thought experiments which circumvented it. The late 1920s resounded to a ding-dong battle between Einstein and Bohr on this issue. The one would propose clever trick after clever trick to beat the Heisenberg relation, whilst the other would show with equal persistence that further thought revealed a flaw in each successive suggestion. In the end the immovable object vanquished the irresistible force. The uncertainty principle survived unscathed.

The final round in the contest involved the time−energy uncertainty principle of equation [4]. Einstein proposed a box full of radiation with a clock-operated shutter, so arranged that it was open for a time Δt, letting out some radiation during this period. There would, therefore be this degree of uncertainty Δt about the actual time at which the radiation was released. Heisenberg laid it down that in this circumstance there should be an uncertainty ΔE about the amount of energy released amounting to at least $\hbar/\Delta t$. Einstein suggested that the box should be weighed before and after the opening of the shutter and the loss of weight determined. The change in mass thus found could be interpreted, via the equation $E = mc^2$ (for which Einstein doubtless had a certain affection), as a change in energy content of the box. In other words, the energy release could be determined exactly, whatever Heisenberg might say to the contrary. This example gave Bohr a sleepless night. In the end, however, he found a most gratifying resolution. In turning the tables on Einstein he used another of the latter's ideas. The acts of weighing involve probing the box's interaction with a gravitational field. That is what weighing is. In his theory of general relativity Einstein had showed that gravitational fields made clocks run slow, an effect called the gravitational red-shift. A detailed analysis of all that was involved in weighing revealed that this induced uncontrollable uncertainties in the rates of slowing of the clock before and after the shutter was opened. Because one did not know how accurately the clock had been ticking during the processes of weighing, one could not know precisely the times at which the movements of the shutter occurred between which the radiation was released. The more accurately one tried to determine the weights (and hence the energy loss) the more uncertain the timing of these instants would become. When all these effects are estimated one obtains a ΔE and Δt which exactly agree with [4].

Although the last word lay with Bohr some people felt there was something fishy about his answer. The appearance of a specific theory of gravitation in the resolution of the problem, even if it were Einstein's own, seemed a trifle odd. I think their anxiety was misplaced. The gravitational red-shift was shown by Einstein, as early as 1911, to follow from the principle of the conservation of energy. That is a principle of sufficient generality to be invoked without disquiet.

After this encounter Einstein gave up his specific attempts to undermine the uncertainty principle. Nevertheless he remained highly sceptical of quantum theory. Chapter 7 contains an account of a penetrating and surprising analysis of some further implications of the theory which Einstein helped to bring to light. Bohr's reaction to this later development was 'Strange but true'; Einstein's 'Too weird to be credible'. The reader can form his own conclusion in due course.

Most of those who resisted the notion of a radical indeterminacy in nature followed a different tack from Einstein. We are not unfamiliar with situations where the best we can do in practice is to assign probabilities, not because events are fundamentally acausal but because their detailed mechanism lies at a level too deep to be accessible to us. An example is provided by Brownian motion. When certain preparations of liquids containing small particles are examined under a microscope the particles are found to joggle around in what appears to be a random fashion. In 1905 (that wonderful year!) Einstein provided the explanation of this phenomenon which had long eluded the understanding of physicists. The particles were light enough to be deflected by the jostling of a swarm of unseen atoms in the fluid as these atoms sped about their business. It was the first occasion on which the atomic constitution of matter had been recognised as producing visible effects. No one then supposed, however, that the atoms were not moving according to strict laws. Their consequence for the tiny particles was only apparently random and unpredictable, due to the hidden atomic motions of which we did not have adequately detailed knowledge.

Could it not be that the same was true in quantum theory? Its apparent unpredictability might be due just to the operation of undisclosed effects – *hidden variables* they were called. On this view, the exact time at which every unstable atom decayed would in fact be fully determined, but by a mechanism of which we were unaware. In a sample of such atoms the different hidden settings of these internal clocks would produce the sort of statistical distribution that quantum theory predicted. But all the time there would be this comforting regularity beneath the surface; the sound of rattling dice would have been banished from fundamental physics. Proponents of this idea were often willing to concede that the hidden variables might be, for some reason, in principle unobservable. We might never be able to take apart the atomic clock to see its wheels and springs, but there would still be the metaphysically soothing thought of its steady ticking. Those who embraced this point of view accepted Heisenberg's uncertainty principle as wholeheartedly as anyone, but for them it was a statement, not of indeterminacy, but of ignorance.

It sounds a beguiling idea, but a little reflection shows that it will not be so straightforward to achieve. After all, quantum phenomena are pretty odd. It is fairly easy to see how such a notion might give the behaviour of randomly decaying atoms. It is rather less easy to see how it would lead to interference effects like those described in Chapter 4. Nevertheless people decided to try it out, till in 1932 it seemed that the *coup de grâce* to such activity had been delivered by the mathematician John von Neumann. He then proved to his own satisfaction that all such endeavours were doomed to failure. According to his calculations such theories would be bound to disagree with quantum mechanics in some of its experimentally verified predictions.

The cat was set among the pigeons in 1952 when, von Neumann notwithstanding, David Bohm constructed a specific example of a hidden variable theory which did agree precisely with conventional quantum mechanics in all its empirical predictions. Of course this made it clear that von Neumann's 'proof' needed reappraisal, a task successfully accomplished by John Bell. The great Hungarian mathematician had not got his sums wrong. His mathematics, as one might have expected, was impeccable. However he had made a harmless-looking technical assumption in setting up the problem which proved to be unduly restrictive. (It related to additivity properties assumed for the way observables depended upon hidden variables.) The reason Bohm could get away with it was that he did not submit his theory to this superfluous requirement.

It is no disrespect to Bohm's achievement to say that his theory is pretty weird all the same. It has to be to get the answers needed. In a hidden variable theory, with everything determinate, each electron in the two-slit experiment of Chapter 4 has to go through a definite slit. To square this with interference phenomena it is necessary to suppose the existence of a strange force acting on the electron as it traverses, say, the top slit, whose nature depends on whether the bottom slit is at that instant open or shut. Such forces, acting instantaneously at one place due to circumstances somewhere else, are called non-local. They are counter-intuitive and many of us feel that they are more disagreeable than the disease of indeterminacy they were invented to cure. In the opinion of many Bohm had jumped out of an indeterminate frying pan into a crackling non-local fire.

This effect is produced by Bohm's adopting an idea of de Broglie's which the latter called the 'pilot wave'. A divorce is decreed between wave and particle which quantum theory had forever joined together. For Bohm there is both a particle (which we see) and a separate wave (which we do not see directly but which creates the strange force on the particle). The wave has no difficulty in going through both slits and provides a 'choppy sea' on which the particle bobs up and down like a cork. The weakness of this approach is the *ad hoc* introduction of this elusive pilot wave; it calls to mind the luminiferous aether of nineteenth-century physics.

For this sort of reason rejuvenated hidden variable theories have not found much acceptance among professional quantum mechanics. They are ingenious but too contrived to be convincing. No one would have constructed them in the form they have if he had not known that at all costs he must, when it comes to experimental predictions, obtain those same results which the statistically interpreted Schrödinger equation seems to produce so economically and naturally. The situation is not unlike that in astronomy after the Copernican revolution. Ptolemy's epicycles could still fit the data. In fact, in some respects at first they did so better than Copernicus's calculations. Yet their forced complexity came to make them seem unconvincing when compared with the eventual attractions of the heliocentric system. There is a deep feeling among those who practise fundamental science – a feeling that has so far always proved reliable – that the way to true understanding is the one that satisfies the canons of economy and elegance; the way which, in a word, is mathematically beautiful. On that score there is no doubt that the principles of quantum theory, as enunciated in Chapter 3, hold the field as the convincing account of the physics of the microworld.

We are used to the idea that quantum mechanics may modify our everyday intuitive notions about the meaning of words like 'position' and 'momentum'. These are the quantities with which physics deals and as the subject ventures into new regimes remote from common experience it must adjust its concepts to match whatever it finds there. No one could object to that. It is altogether more surprising to find that quantum mechanics also causes us to re-evaluate words like 'and' and 'or'.

You tell me that Bill is at home and either he is drunk or he is sober. Accordingly I expect that either I shall find Bill at home drunk or I shall find him at home sober. If we wanted to be self-consciously intellectual about it we could say that this little argument uses the logical principle of the distributive law. That is rather a ponderous way of saying that no one in his right mind doubts the truth of my assertion about the states in which Bill will be found, because there are two of them and he must be in one or the other.

If we replace Bill by an electron something odd happens. You will remember that electrons have a quantity called spin whose component in any assigned direction can take only two values, 'up' or 'down' (p. 22). That statement is true for the components in whatever directions we are pleased to choose. Let us apply it to two mutually perpendicular directions, the x and z axes let us say. We will call the corresponding spin components s_x and s_z respectively. Let us suppose that we know that s_z is 'up'. Then it is certainly true to say that

s_z is 'up' and s_x is either 'up' or 'down'.

A classical logician, like Aristotle or the man in the street, would go on to deduce from this that

either s_z is 'up' and s_x is 'up' *or* s_z is 'up' and s_x is 'down'.

That, however, is wrong. The reason is that the observables s_x and s_z do not commute with each other and it is therefore impossible to have states in which they both take assigned values, like both being 'up', or one 'up' and one 'down', as the second statement would imply if it were correct. In fact, if s_z is 'up' then the electron is in a state which is an even-handed superposition of the states in which s_x is 'up' and s_x is 'down'. That is a middle term undreamed of by the founders of classical logic. It indicates the possibility, indeed the need, to generalise logical structures to take account of such peculiarities. The resulting *quantum logic* was initiated by the mathematicians John von Neumann and Garret Birkhoff, using mathematical constructs called lattices.

It is a curious little story. I suppose that logicians and physicists normally consider themselves to be poles apart. The physicists, whilst they endeavour to be rational, are apt to consider the formalities of logic to be arid and deserving of the prefix 'chop'. Logicians, one would have imagined, dwell in mental ivory towers, austerely untroubled by what goes on in laboratories. Yet both had lessons to learn from the behaviour of electrons. In the end, the way the world is affects us all.

Chapter 6

Fixing it

The Schrödinger equation is as nice a differential equation as any classical physicist could wish to see. All is smoothness and continuity. It is only when we try to extract some information by means of a measurement that the traumatic discontinuity of the collapse of the wavepacket takes place. Here is the unique point at which the fitful indeterminacy of quantum theory makes itself felt. Here, in consequence, is the source of perplexity and debate.

Measurement involves an intervention by our everyday world into the quantum world. Our macroscopic world is precise; it is the domain of classical Newtonian physics. The microscopic quantum world is imprecise; it is the domain of Heisenberg uncertainty. How do these two worlds interlock? In particular, how does it come about that the imprecise quantum world yields a precise answer when it is experimentally interrogated? What fixes the result of a particular experiment when the theory is only able to assign probabilities for a variety of possible outcomes?

Because we do not have direct access to the microworld, any measurement involves a chain of amplification by which the state of affairs on the very small scale is made to manifest a corresponding signal in the everyday world of the laboratory. In the time-honoured language of thought experiments, the pointer moves across the scale to a mark saying 'here'. Let us consider such a chain of consequence, spanning small to large.

Suppose we are trying to determine the spin of an electron along a particular direction, to find out whether it is 'up' or 'down'. The standard technique is to use a Stern–Gerlach experiment. The electron is made to pass through a magnetic field which deflects it in different directions according to the different orientations of the spin. If the spin is 'up' the electron goes off one way, if it is 'down' the electron goes off the other way, as in the figure. We can place some detecting devices, photographic plates or Geiger counters, say,

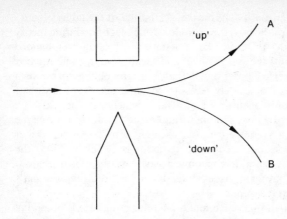

which will register whether the electron has arrived at A or B. If we analyse such an experiment what we obtain is a chain of correlations:

if the electron's spin is 'up', then it will be deflected to A, and then the Geiger counter at A will click
or
if the electron's spin is 'down', then it will be deflected to B, and then the Geiger counter at B will click.

We can extend the chain to take into account the presence of an observer who 'then hears the counter at A(B) click'.

You see what is happening? The buck of decision is being passed down the chain of consequence. At no stage do we say *now* this happens (the counter at A actually clicks) but always *then* this happens (if ... then the counter at A clicks). All that is being asserted is a correlation. Yet we know that the matter does get settled somehow. Experiments actually yield results. The question is, where along the chain does the collapse of the wavepacket take place? Where do we shrug off the 'I will if you will' tale of correlation and reach an actual decision about what has occurred on this occasion?

The need to settle this point is emphasised by the stories of two of the cast list in quantum mechanical folklore: Schrödinger's cat and Wigner's friend.

The unfortunate animal in question is incarcerated in a closed box which also contains a radioactive atom with a 50–50 chance of decaying in the next hour, emitting a γ-ray in the process. If this emission takes place it triggers the breaking of a vial of poison gas which instantaneously kills the cat. At the end of the hour, before I lift off the lid of the box, the orthodox principles of quantum theory bid me consider the cat to be in a state which is an even-handed superposition of the states 'alive' and 'dead'. On opening the box the wavepacket collapses and I find either a cooling corpse or a frisking feline. It is scarcely necessary to emphasise the absurdity of the proposition that this state of affairs, whichever it is, has been brought about by my action in lifting the lid. It must surely be the case that the cat is competent to act as observer of its own survival or demise and does not need me to settle the issue for it. Along the chain of consequence, from atomic decay to my act of observation, things must have got fixed at least by the time that the cat's experience entered into it.

Wigner's friend serves to make a similar point. He is observing the electron being deflected by the magnetic field in the Stern–Gerlach experiment and listening to the Geiger counters to see whether it is the one at A or the one at B which clicks. Wigner knows *a priori* that the probabilities of these two events are equal. The electron beam is unpolarised (as we say), so that it is an even-handed superposition of 'up' and 'down'. On a particular occasion Wigner asks his friend which counter clicked. The friend replies that it was A. For Wigner the electron's wavepacket then collapses into the state in which the spin is definitely 'up'. Yet this state of affairs has surely not been brought about by Wigner's intervention. If he asks his friend 'What did you hear before I asked you?' his friend will reply with some impatience 'I told you before, the counter at A clicked'. Wigner has to take his friend's claim to experience as seriously as we take Wigner's. At least by the time that the friend heard the click it must have been settled on that occasion that the electron's spin was 'up'.

Schrödinger's cat and Wigner's friend are links in the chain from microscopic object to my knowing the result of a measurement made upon that object. The reasonable interpretation of their experience seems to demand that by the time the chain had reached them the actual result of the observation had become fixed. The cat knows whether it is dead; Wigner's friend knows what he heard. In other words, the intervention of consciousness in either a developed form (Wigner's friend) or at a more rudimentary level (Schrödinger's cat) seems to provide the latest link in the chain at which the matter could be settled. Of course, it might be the case that it had been settled long before that.

Four different ways have been suggested in which one might seek a resolution of the problem of the collapse of the wavepacket. None of them is free from difficulty.

The first one tries to cut the chain. It says that the wavefunction is not a description of a physical system but simply a description of my knowledge of it. It does not worry where that knowledge comes from. There is, of course, no problem about a state of knowledge suffering a sharp change. I have no idea which football club won the FA Cup in 1981. Consulting a reference book 'collapses my soccer information wavepacket' into the unique state 'Tottenham Hotspur'. I do not need to agonise about that sort of discontinuity of experience. It is clear enough what has happened. I was ignorant and then I was enlightened. The change is in my mind.

It sounds seductively simple to say that it is all in the mind. Bishop Berkeley, in the eighteenth century, encouraged us to contemplate just such a step. He would, however, be an ill-chosen ally for quantum mechanics, for a retreat into the idealist castle would be a highly injudicious move. It would have the effect of demoting physics to a branch of psychology. The subject cannot allow itself to suffer such a fate. The knowledge with which it deals originates outside us; the concept of nature standing over against us is not lightly to be abandoned. Without it the point of physics would be lost, the authentic experience of discovery denied. (Nevertheless I must admit that this point of view counts among its adherents Max Born, who was one of the founding fathers of the subject.)

We must, therefore, take seriously the chain of correlated consequence linking electron to observer. The second possibility is to say that things get fixed along that chain once the objects with which we have to deal have become 'large'. In essence this is the answer of what is called the Copenhagen Interpretation. Bohr and his friends hammered out an approach to the understanding of quantum theory which hardened into a rigid orthodoxy prescribed for the faithful. An important part of their creed was the existence of classical measuring instruments. The pointers and other registration devices incorporated in these pieces of apparatus act with Newtonian dependability. There is no doubt at all whether the reading says 'up' or 'down'. The Copenhagen school therefore says that it is at this stage that results get fixed and knowledge is established. The conscious observer can then take note if he wishes to do so. To be sure it is a little difficult to be certain how big 'large' has to be before this behaviour sets in – indeed it was emphasised by Bohr and his friends that the line could be drawn at different places along the chain – but somewhere this classical behaviour had to set in to make measurement possible.

The proposal has a certain attraction. If we go into a laboratory we shall find it littered with devices of just this character, capable of affording an unambiguous answer to an experimental enquiry. The problem is to understand how this undoubted fact finds a consistent place within quantum theory. You can see the difficulty if you recall our discussion of Heisenberg's γ-ray microscope in Chapter 5. At first sight the microscope would seem to be just the sort of classical measuring instrument that Bohr and his friends had in mind. But you will remember that we ran into trouble if we pressed that point to its logical conclusion. If the microscope were truly classical, untroubled by Heisenberg uncertainty, we could then determine its recoil on absorbing the photon and use that knowledge to circumvent the uncertainty relation for the electron itself. It seemed that our quantum mechanics had to be thorough going, even if we could perceive, as we did at the end of Chapter 4, circumstances in which classical calculations would provide an acceptable approximation. This latter fact simply means that as we go along the chain of correlated consequences to larger and larger systems the links in the chain become tighter and tighter, less and less subject to quantum mechanical 'creakiness'. It does not explain how the intervention of these (nearly) classical systems *chooses* one chain rather than another as the physical occurrence on a particular occasion. By assigning such a determining role to classical instruments Bohr divided the world into two, the uncertain quantum world and the objectifying world of measuring instruments. The example of the γ-ray microscope shows that it is far from clear how these two worlds could be juxtaposed without subverting and contradicting the very quantum theory they were invented to explain.

The same point can be made more formally by thinking about the Schrödinger equation. If we apply the equation simply to the electron by itself then the act of the microscope in determining the electron's position has to be represented in a *deus ex machina* way as an external intervention bringing about the discontinuous collapse of the wavepacket. However there is, presumably, a gigantic Schrödinger equation which describes both microscope and electron. It is, of course, of a complexity far beyond our power to solve. However in principle it exists and describes the evolution in time of the joint system of electron plus measuring apparatus. From this point of view everything changes smoothly. This unruffled progression fits perfectly the sort of chain of *correlations* of consequences which we disentangled at the start of this chapter but it does not seem to offer the prospect of the decisive determination of a particular *result*.

The third suggestion seizes on the role of the conscious observer to provide the resolution of our difficulty. Here, at the end of the chain, is the intervention of an agency of a manifestly different character from all that has preceded it. It may be hard to tell 'large' from 'small', or to bring classical and quantum objects into consistent association, but it seems far less perplexing to find a difference between the mental and the physical and so to attribute a special property to the interface of consciousness at which they meet each other. Consciousness does have a unique character about it, a character which we know from direct experience and which we can reasonably attribute by analogy to other human beings and, perhaps, to cats. We said that this was the last stage before which any unfixedness could be tolerated. Maybe it is the actual point at which things are settled. It will not surprise you to learn that Eugene Wigner has been a leading proponent of this view. It is not so drastic a solution as our first all-in-the-mind theory which attributes to the wavefunction a significance purely in terms of human knowledge, for this third view takes the external world much more seriously as the origin of the chain of correlated consequence.

Accordingly we can ask its supporters the same question we addressed to our friends from Copenhagen. How does the postulated discontinuity appear when we consider the enormous Schrödinger equation embracing the totality of all that is involved in the measurement process? This will be a truly gigantic equation in this instance, for it will have to describe not only the electron and the microscope but also the observer peering down the latter's eyepiece. Will there not be a continuity of evolution implied, in contradiction to our postulated discontinuous collapse?

Presumably the answer will be 'yes' if we accept the premise of the question. This has led supporters of this particular point of view to deny the premise. It involves an immense extrapolation of our actual knowledge to assert that the Schrödinger equation would furnish an adequate description of a system sufficiently complex as to be able to sustain consciousness. Of course the thrust of the programme of molecular biology is to extend contemporary physics into biology in just such a way. Impressive as its successes have been in such matters as the unravelling of the genetic code, nevertheless they represent only an infinitesimal step towards a complete account in conventional physical terms of even the simplest animal brain. To say this is not necessarily to be an obscurantist hankerer after vitalism, a wistful adherent of a dualist view of 'the ghost in the machine'. It is just to be cautious about carrying ideas far beyond the regimes in which they have currently been tested. It would be possible to suppose, for instance, that the true Schrödinger-like equation, whatever it is, involves non-linearities which enforce a violation of the superposition principle for very complex systems, whilst essentially preserving it for simple systems to which we have actually applied quantum mechanics. Indeed Wigner has formulated a proposal along exactly these lines. In principle this could give some rational insight, in terms of an extended scientific model, into how conscious observers could produce wavepacket collapse.

Nevertheless there is something very unattractive about this particular suggestion. It is astonishingly anthropocentric or, at best, biocentric. Are we to suppose that in the thousands of millions of years before conscious life emerged in the world – and still today in those extensive parts of the universe where no conscious life has yet developed – no wavepacket has ever collapsed, no atom for certain decayed? That quantum mechanics as we know it is a biologically induced phenomenon? That photographic plates stored away uninspected at the end of an experiment only acquire a definite image when someone opens the drawer to have a look at them? It takes a bit of swallowing.

The dilemma of those who evoke consciousness as the basis of phenomena was succinctly stated by Ronnie Knox in his limerick on idealism:

There once was a man who said 'God
Must think it exceedingly odd
If he finds that this tree
Continues to be
When there's no one about in the Quad.'

An anonymous author provided an answer along lines approved by Bishop Berkeley:

Dear Sir, Your astonishment's odd;
I am always about in the Quad.
And that's why the tree
Will continue to be,
Since observed by Yours faithfully, God.

Such a riposte would not, however, be available to a defender of the interpretation of quantum mechanical measurement presently under consideration, supposing him to wish to avail himself of it. Divine reduction of wavepackets would be an overkill, since it would operate everywhere and always, forcing the electron each time to go through a definite slit. The point about measurement is that it only occurs spasmodically. [This observation is in accord with the classic theological understanding of creation, which sees God as the ground and support of all that is (in our terms, the guarantor of the Schrödinger equation) but not as an object among objects (no collapser of wavepackets).]

It might seem that we have now exhausted all possible ways of trying to comprehend quantum mechanical measurement. There remains, however, one further suggestion which easily exceeds all others in its bizarre quality. This is the many-worlds interpretation proposed in 1957 by Hugh Everett III. It arose in the context of the thinking of people whose job it is to think big. They were cosmologists wrestling to apply quantum mechanics to Einstein's general theory of relativity, for which the natural system for consideration is the universe itself. Here is a case where talk of classical measuring apparatus, or the role of an external observer, is simply squeezed out. There is no room for their separate existence in the all-embracing picture of the cosmos. The quantum mechanical formalism itself is left as the sole source of insight. A daring proposal was made to reconcile the continuity of the Schrödinger equation with the discontinuity of empirical experience. It was supposed that, in every circumstance in which there is a choice of experimental outcome, in fact each possibility is realised. The world at that instant splits up into many worlds, in each of which one of the possible results of the measurement is the one that actually occurs. Thus for Schrödinger's cat there is one world in which it lives and another world in which it dies. These worlds are, so to speak, alongside each other but incapable of communicating with each other in any way. This latter point is supposed to explain our feeling that we experience a continuity of existence for ourselves. The cat who lives is unaware of the cat who dies. All the time I am being repeatedly cloned as copies of myself multiply to pursue their separate lives in the many worlds into which my world is continually splitting. Each of these *alter egos* is unaware of the others, so that he is untroubled by his derivation from the common ancestor he unwittingly shares with the many other Polkinghornes who branched off in the course of some quantum mechanical act of measurement.

It is enough to make poor William of Occam turn in his grave. Entities are being multiplied with incredible profusion. Such prodigality makes little appeal to professional scientists, whose instincts are to seek for a tight and economic understanding of the world. Very few of them indeed have espoused the Everett interpretation. It has, however, been more popular with what one might call the 'Gee-whizz' school of science popularisers, always out to stun the public with the weirdness of what they have to offer. 'Are you on an aircraft about to crash?' they ask. 'Don't worry, there are other worlds in which the quantum fluctuations have gone along a different path which will prevent the crash. Those carbon copies of yourself which inhabit those worlds will live on.' Personally I should find that cold comfort. Reality is not to be trifled with and sliced up in this way.

It is a curious tale. All over the world measurements are continually being made on quantum mechanical systems. The theory triumphantly predicts, within its probabilistic limits, what their outcomes will be. It is all a great success. Yet we do not understand what is going on. Does the fixity on a particular occasion set in as a purely mental act of knowledge? At a transition from small to large physical systems? At the interface of matter and mind that we call consciousness? In one of the many subsequent worlds into which the universe has divided itself? Of the suggestions that we have canvassed the one that seems to be the most promising is that which points to the emergence of classical measuring instruments from a quantum mechanical substrate, even if the mode and consistency of this happening is harder to comprehend than Copenhagen orthodoxy is willing to recognise.

It is a puzzle with a familiar ring to it. Our knowledge of the world is organised into subjects which form a hierarchy, graded by the degree of complexity of the systems which they treat as basic. At the bottom is physics, the most reductionist of all; above it there is biology; then psychology and anthropology; sociology; and (some of us would wish to add) theology. At each level in this hierarchy we find both a dependence on those subjects which lie below and also a claim to a degree of autonomy for the concepts which are specific to that level. How these claims and dependencies are reconciled, how the views of the different disciplines fit consistently together to give a total picture of the one world of experience is a problem both vast and baffling. As someone who is both a physicist and a priest I am aware of some of the puzzles involved in teasing out the relation of the scientific and religious views of the world. To take another example, there are many biologists who, despite the acknowledged successes of molecular biology, do not believe that their subject is just a complex corollary to physics. My instinct is to believe that those who defend the autonomy of their particular 'level' are right to do so but I also think that our present state of knowledge is insufficient to permit us to understand satisfactorily how this level autonomy comes about. We no more understand how biology emerges from physics than we understand how classical measuring apparatus emerges from quantum mechanics. But it is at least interesting that there seems to be this level shift problem embedded within physics itself.

Chapter 7

Real togetherness

Einstein licked his wounds after his long drawn out battle with Bohr about the uncertainty principle. The great man was bloody but unbowed. In 1935, with two young collaborators, Boris Podolsky and Nathan Rosen, he returned to the attack. They produced an analysis of some of the consequences of conventional quantum mechanics which they felt would convince people of the unsatisfactory nature of that theory. We shall give an account of the Einstein–Podolsky–Rosen paradox in a simplified form due to David Bohm. I will try to make it as easy as possible to understand but it will require a degree of concentration to follow the argument. After all, if it did not occur to Bohr, Heisenberg and Schrödinger in the twenties it cannot be trivially self-evident.

One can readily obtain pairs of protons in what is known as a singlet state. This means that their spins are guaranteed to cancel each other out to give a total spin of zero. Suppose two such protons, A and B, are allowed to separate widely apart. We now measure the spin of A along some direction, let us say the z-axis. We find that this quantity, $s_z^{(A)}$, is 'up'. Because the spins have to cancel it follows that B must be in a state where its component of spin along the z-axis, $s_z^{(B)}$, is definitely 'down'. If instead we had measured A's spin in the x direction and found it to be 'up' then, of course, by a similar argument B would have been in a state with $s_x^{(B)}$ definitely 'down'. The general rules of quantum mechanics tell us that this latter state is an even-handed superposition of the states in which $s_z^{(B)}$ is 'up' and $s_z^{(B)}$ is 'down'.

You can see that something odd is happening. Measuring a spin component of A has an instantaneous effect upon B, causing its spin wavefunction to collapse into the state with the opposite spin component. Moreover, different and indeed contradictory things happen to B according to which component of A it is that I choose to measure. If it was $s_z^{(A)}$, then B is forced into a state with a definite

$s_z^{(B)}$, whilst if it was $s_x^{(A)}$ that I measured then B is then in a state which is an equal superposition of the states in which $s_z^{(B)}$ is 'up' and 'down'. Thus an instantaneous influence propagates from A to B whose effect depends radically upon exactly what it is that I measure on A.

It is certainly counter-intuitive to suggest that widely separated systems retain such a degree of influence upon each other. It is instructive to compare the quantum situation with the system as classical physics would describe it. Newton would assign equal and opposite spins to A and B. Their values would be fixed at the moment of separation even if they might not be known to an observer until he had actually measured them. No such act of measurement could change what those spins already were; it could only bring to light a state of affairs which had existed from the moment of separation. Once I know that A's spin has a particular z-component then I know that B has just the opposite. In no way, however, has the act of measurement on A brought about a change in B. It was always that way. Classically, once A and B are apart they are truly separated. Quantum mechanically this is not the case. Since different measurements on A produce incompatible consequences for B, the act of measurement must induce a new state for B, not just reveal its previously existing state. Once again we can see how peculiar a hidden variable theory would have to be to reproduce all the consequences of quantum mechanics (recall the discussion of p. 57). It would need non-local forces which instantaneously transmit effects from A to B.

There could be no doubt that EPR, as we shall acronymically refer to them, had drawn attention to a very striking feature of quantum mechanics. Was it however so paradoxical as to have dealt a death blow to the theory?

Bohr as usual was imperturbable. The Copenhagen school had made a special point of emphasising that one ought never to think of quantum mechanical systems without also annexing to them the array of classical measuring instruments with which it was proposed to make the observations. It was, in their view, a package deal. Change the observations you were going to make and you had a new situation, even if the system to be observed remained the same. This was their way of making the celebrated assertion that quantum mechanics does not allow a separation between observer and the system observed.

Measuring the z-component of the spin of A involved the use of one sort of apparatus; measuring the x-component involved another. The two experimental set-ups were different, incompatible, and so could not act together. In a word, they were complementary. Why then be surprised that they led to incompatible consequences, even if that included different effects upon the distant system B?

This reply illustrates the strengths and weaknesses of the somewhat positivist approach of Bohr and his friends, with their emphasis on classical measuring apparatus. It enabled them to shrug off EPR, but at the cost, one might think, of refusing to face the issue. There is a way of proceeding in conceptual matters whose method is to define away any inconvenient difficulty. All the really tricky questions are declared meaningless, despite the fact that they are sufficiently well comprehended to give rise to perplexity. On the EPR paradox it seemed that the Copenhagen school had achieved just such a Pyrrhic victory.

Einstein, of course, was a realist through and through. His analysis of the situation ran along different lines. It was based upon two principles:

1. The Reality Principle. The question of the reality of the world outside us is acknowledged to be a tricky one. We do not attribute reality to all the objects of our apparent perception. Hallucinations and dreams are to be discriminated from sober waking experience. An important clue to this act of discrimination has always been found in the existence of regularities, especially where they make prediction possible. Where this is the case men have thought that they were in touch with reality. Clearly quantum mechanics, with its irreducible degree of disturbance in making a measurement and the consequent uncertainty principle limiting knowledge, has not eased this problem. (More of this in Chapter 8). Einstein and his collaborators had, therefore, to be very careful how they defined what they meant by physical reality. They appeared to have been suitably cautious in offering the following definition: *If, without in any way disturbing a system, we can predict with certainty (i.e. with probability equal to unity) the value of a physical quantity, then there exists an element of physical reality corresponding to that quantity.*

To this they tactitly added a second principle. It seemed at the time too obvious for extensive formulation, but had they spelt it out it would have amounted to something like

2. *The Locality Principle. If two systems have been for a period of time in dynamical isolation from each other, then a measurement on the first system can produce no real change in the second.*
It sounds perfectly reasonable, but you will perceive that here is the nub of the matter.

Armed with these two principles Einstein, Podolsky and Rosen set to work to demolish quantum mechanics. (I translate their argument into the form appropriate to the discussion of Bohm's simpler version of the EPR experiment.) A measurement of $s_z^{(A)}$, would enable us to predict with certainty the value of $s_z^{(B)}$, namely the opposite one. Equally the measurement of $s_x^{(A)}$ would enable us to predict the value of $s_x^{(B)}$ with certainty. On the basis of our stated principles we can therefore say that $s_x^{(B)}$ and $s_z^{(B)}$ are both real properties of the proton B. However quantum mechanics does not permit them to enjoy this status simultaneously, since they correspond to operators which do not commute with each other (p. 28). Rather than calling this a paradox (as others did) the authors preferred to think of it as an indication of the *incompleteness* of quantum theory. A complete theory would need to accommodate every element of physical reality, something which quantum mechanics in their view failed to do. The EPR paper concludes with the words

While we have thus shown that the wavefunction does not provide a complete description of physical reality, we have left open the question of whether or not such a description exists. We believe, however, that such a theory is possible.

The wistful longing after a determinate hidden variable theory clearly remained unassuaged.

EPR had won the battle on their own terms but, like Bohr in his very different way, they had defined the rules so as to assure themselves of victory. The crucial issue is the locality principle. That celebrated sage, the man on the Clapham omnibus, would no doubt readily assent to it. The quantum world, however, has to be allowed to construct its own notions of (uncommon) sense. The natural way to interpret the EPR experiment is not that it shows up the incompleteness of quantum theory but that it manifests the falsity of naive locality. For quantum systems it seems that once they have met there is never true parting. They enjoy a lasting degree of real togetherness.

However illuminating thought experiments may be, nothing carries conviction like an actual series of measurements made in a real-life laboratory. The problem of the nature of locality, raised by EPR, clearly demanded some form of empirical investigation. To bring it into a form suitable for testing involves a modest degree of reformulation. Our basic principles will be threefold:

1. *Reality: Regularity of phenomena is due to an underlying physical reality.* This is a more general statement than 1. above, but consonant with the spirit of our previous discussion which alleged that regularity should be the touchstone for telling reality from illusion.

2. *Locality.* This is what we are particularly keen to probe. We can assume it in exactly the form stated above or we can relax it to the extent of saying that any influence of A upon B must not propagate between them faster than the velocity of light. (I will say a little more about relativity at the end of the chapter.)

3. *Induction: It is possible to reach conclusions valid for all systems of a given type from a consistent set of observations on a large sample of systems of that type.*
   Whatever may be the logical difficulties of a principle of induction (and there has, of course, been much written on the subject from David Hume to Karl Popper), as a methodological strategy it is essential for science. Since we can never investigate all protons, any general statement about them whatsoever must depend upon a principle of this sort.

These three principles together define a theory of a type which Bernard D'Espagnat calls *locally realistic*. From them an empirically testable consequence can be drawn which is in contradiction with the predictions of quantum mechanics. The route to this prediction is a little roundabout, but the resulting inequality (on p. 76) gives the experimenter the chance to decide the issue between locally realistic theories and conventional quantum mechanics.

We set up an apparatus in which pairs of protons, A and B, are produced in a singlet state. The protons separate and each passes through an analyser which can determine its spin along one of three directions (not necessarily mutually perpendicular) which we label

α, β and γ. If the spin is 'up' in the direction $\alpha(\beta,\gamma)$ we label the result α_+ (β_+, γ_+); if it is 'down' we label the result $\alpha_-(\beta_-, \gamma_-)$. The measurements on the spin components of A and B are either made simultaneously or at least in such a way that no influence moving with the velocity of light can pass from one measurement to the other. The apparatus can be arranged in such a way that sometimes we measure the same component for both A and B (both α, for example) and sometimes we measure different components (for example, β for A and γ for B).

Every time we measure two α components we find that they cancel out. If A is α_+ then B is α_-, or vice versa. Principles (1) and (2) lead us to interpret this regular correlation as an indication of the existence of a local reality; that is, being in a state α_+ or α_- is a real property of A (or B). Invoking (3) leads us to assert that this is a real property of A (or B) even in the circumstance that we make no measurement of the α component. (This is a conclusion that quantum mechanics denies, of course.) Similarly we deduce physical reality for the states $\beta_\pm$, $\gamma_\pm$. Thus a proton state, on this view, is labelled by three choices of $+$ or $-$, to indicate the values of the corresponding spin components in the directions α, β, γ. There are eight possibilities in all: $(\alpha_+ \beta_+ \gamma_+)$, $(\alpha_- \beta_+ \gamma_+)$, ..., $(\alpha_- \beta_- \gamma_-)$. We cannot determine experimentally all these quantities for a given proton but we can fix two of them. Suppose I measure α for A and find $+$, and β for B and find $+$ also. Because of the singlet condition I know that β_+ for B implies β_- for A, so that A certainly has α_+, β_- and it must be in one of the two states $(\alpha_+ \beta_- \gamma_+)$ or $(\alpha_+ \beta_- \gamma_-)$. From this sort of consideration, and some combinatorial mathematics which I will not inflict upon you here, John Bell deduced in 1964 an experimentally testable consequence (see Appendix, A9).

An experiment of the type we are considering is conducted with a large batch of proton pairs. The spins of the two protons are measured along axes which for each proton on each occasion are selected at random from the possible choices α, β, γ. Sometimes both protons have their spins measured in the same directions (both α, say), sometimes they are in different directions (one β, one γ, say). After a long series of such measurements I add up the number of instances in which a particular combination of results has occurred. For example the number of times that one or other proton has α_+ and the other then has β_+ would be a number that I shall call $n(\alpha_+, \beta_+)$. Numbers $n(\beta_+, \gamma_+)$, $n(\alpha_+, \gamma_+)$, etc., are similarly defined. Bell showed

$$n(\alpha_+, \beta_+) \leqslant n(\alpha_+, \gamma_+) + n(\beta_+, \gamma_+),$$ [1]

or, in words, the sum of $n(\alpha_+, \gamma_+)$ and $n(\beta_+, \gamma_+)$ is always greater than or equal to $n(\alpha_+, \beta_+)$. Here is a testable prediction, all the more interesting for the fact that for appropriately chosen orientations of the axes α, β, γ, one can show that quantum mechanics leads to a violation of the Bell inequality.

Although this is a practical proposal, the way in which we have described it is not free from idealised elements. We assume, for instance, 100 per cent efficiency for the measuring devices, which is better than technology can deliver. All that has to be taken care of in designing experiments to probe local reality. These experiments are difficult, but the interest in the outcome is sufficiently high for a number of careful experimental groups to have been encouraged to make the attempt. Almost all experiments use photons rather than protons. (A similar analysis applies.) There was some disagreement among the results of early experiments but recently Aspect and his collaborators in Paris have performed a beautiful and convincing investigation reporting results in disagreement with the inequality. It thus seems that the evidence is clearly in favour of quantum mechanics and against local reality.

The violation of local reality logically requires, of course, only the denial of one of its foundation principles, (1), (2) or (3). Most people would feel that it is the locality principle (2) which is the one to be sacrificed.

We are back again to the idea that quantum systems exhibit an unexpected degree of togetherness. Mere spatial separation does not divide them from each other. It is a particularly surprising conclusion for so reductionist a subject as physics. After all, elementary particle physics is always trying to split things up into smaller and smaller constituents with a view to treating them independently of each other. I do not think that we have yet succeeded in taking in fully what quantum mechanical non-locality implies about the nature of the world.

A final paragraph for the benefit of a reader with some knowledge of special relativity. I have several times spoken of an 'instantaneous' effect upon B as a result of a measurement made on A. Is this not in flagrant contradiction to Einstein's rule that signals do not travel faster than the velocity of light? The matter is, in fact, not quite so straightforward. What Einstein requires is that no messages should be transmitted which would, for example, permit clocks at A and B to be synchronised simultaneously. However the collapse of the wavepacket does not seem to carry information of this kind so that there does not appear to be a prima facie case of contradiction.

Chapter 8

What does it mean?

If you thought that science was invariably characterised by clarity of vision you may have found the quantum world unexpectedly murky. The brilliant calculational successes can seem a little brittle when they are coupled with the degree of conceptual confusion still present. What does quantum mechanics really have to say about the nature of the physical world? We have to confess that we are at a stage of understanding where any answer must be to some extent tentative.

There are two possible lines of attack, of either positivist or realist tone. The positivist approach lays stress on perceptions which can be inter-subjectively agreed; its tests of meaning and truth rely upon the specification of observable procedures. Such content as it attaches to physical reality is to be interpreted in terms of the experience of an observer; its quest is for the harmonisation of such experience. The world it presents is populated by pointers on scales and marks on photographic plates. The observer disposes the apparatus and takes the readings. His is the central role.

It is a curiously unreal state of affairs, a world that none of us lives in outside the study. As Feynman has said, we are asked to believe that the historian who makes a statement about Napoleon simply means that there are books in libraries which make assertions similar to his own. There is no past; there are only sources.

The approach to quantum theory which is most positivist in its outlook is that of the Copenhagen school, with its emphasis on the role of classical measuring instruments. The very choice by an observer of the disposition of these instruments is considered fundamental to the nature of what is going on, as we saw when we considered Bohr's understanding of the EPR experiment (p. 72). In his public utterances Bohr was always very cautious about committing himself to what it is that actually *is*. He never made ontological pronouncements. He preferred to think of quantum theory as a calculational procedure, writing that†

The entire formalism is to be regarded as a tool for deriving predictions, of definite or statistical character, as regards information obtainable under experimental conditions described in classical terms.

However, in private conversation with a friend, Aage Petersen, he went further and declared

There is no quantum world. There is only an abstract quantum physical description. It is wrong to think that the task of physics is to find out how nature *is*. Physics concerns what we can say about nature.

The austerity of the positivist programme may at first sight seem highly scientific, with its rigid adherence to what can be measured and its banishing of all that is not the immediate fruit of experience. In fact, it is more characteristic of natural history than of science to be content with the patient observation of phenomena without seeking an underlying understanding of the originating reality. If in the end science is just about the harmonious reconciliation of the behaviour of laboratory apparatus, it is hard to see why it is worth the expenditure of effort involved. I have never known anyone working in fundamental science who was not motivated by the desire to understand the way the world is. The point has been well put by Bernard D'Espagnat in his book *Conceptual Foundations of Quantum Mechanics*. Speaking of the activities of elementary particle physicists, he writes that

whereas the activity appears essential as long as we believe in the independent existence of fundamental laws that we can still hope to know better, it loses practically its whole motivation as soon as we believe that the sole objective of the scientists is to make their impressions mutually consistent. These impressions are not of a kind that occur in our daily life. They are extremely special, are produced at great cost, and it is doubtful that the mere pleasure their harmony gives to a selected happy few is worth such large public expenditure.

Or, I would add, were the positivist view correct, the dedication and toil of those involved. Let us see then whether realism can offer a more fruitful alternative.

  A realist approach lays stress on the belief that the world has an

—

† This and the subsequent quotation are culled from the plentiful stock provided by Max Jammer in *The Philosophy of Quantum Mechanics*.

existence independent of any observer; that it stands over against us as an entity in its own right. The content it attaches to physical reality makes the natural world autonomous; its quest is to determine what *is*. The world it presents is populated by entities such as electrons and quarks. By observation we can probe this world and attempt to discern the laws which regulate it. The observer has to submit himself to the way things are.

Unquestionably such a view corresponds to the motivation of science and the way that scientists talk about their discoveries. However we have seen that quantum theory places considerable restraint on a plain man's objectivist view of the natural world. The states of motion permitted to its particles cannot be characterised by straightforward assignment of position and momentum. These states of motion are subject to instantaneous change through the act of measurement, in a process for which we cannot claim to have discovered an exhaustive and convincing interpretation. Despite the atomising tendency of fundamental physics, with its motto 'divide and rule', we have found that the EPR experiment points to a surprisingly integrationist view of the relationship of systems which have once interacted with each other, however widely they may subsequently separate. Even the classical certainties of laboratory apparatus present us with a mystery when we recognise that we do not fully comprehend how they arise from the quantum substrate of which they are composed.

Perhaps some modest help can be obtained from reconsidering how it is that practitioners of quantum mechanics actually go about their trade. We tried in Chapter 3 to give an account of their procedures. The wavefunction (or state vector, in the more abstract language) plays a fundamental role. When a quantum mechanic thinks about what an electron is 'doing' he is thinking about what sort of wavefunction is associated with it. Now, we would all agree that the wavefunction is not a directly physical object in the sense that a billiard ball is a physical object. Indeed the rules of quantum theory do not even allot a unique wavefunction to a given state of motion, since multiplying the wavefunction by a certain factor (technically, a factor of modulus unity) does not change any physical consequence. However it is also difficult to think of a wavefunction as a *mere* calculational device, in the way that Bohr's words quoted above suggested. Its status is somehow intermediate. Heisenberg had something helpful to say on this question. He was a loyal member of the Copenhagen school but he displayed a greater flexibility than most in the expression of his understanding. He once wrote

In the experiments about atomic events we have to do with things and facts, with phenomena which are just as real as any phenomena in daily life. But the atoms or elementary particles are not as real; they form a world of potentialities or possibilities rather than one of things or facts.

Let us take the second half of his concluding sentence first. Here Heisenberg refers to a notion to which he often had recourse, that quantum mechanics had revived Aristotle's old idea of *potentia*. Quantum objects do not act as carriers of classical quantities such as position and momentum (their wavefunctions are not usually eigenstates of these observables) but they do carry the potentiality for such quantities (their wavefunctions are always superpositions of such eigenstates). It is as though the programme of Galileo and Locke, which involved discarding secondary qualities (colour, taste, etc.) in favour of primary qualities (the quantities of classical mechanics), had been carried a stage further and these primary qualities had themselves become secondary to the property of *potentia* in which they all lay latent.

The first part of Heisenberg's concluding sentence shows that he felt that such a view fell short of attributing reality to elementary particles. Perhaps that was too grudging a position. It may be that we can seek help from what to some of my readers might seem an unlikely source. A theologian, Eric Mascall, in his book *Christian Theology and Natural Science*, wrote about such problems that

the point is that, although a physicist knows the objective world only through the mediation of sensation, the essential character of the objective world is not sensibility but intelligibility. Its objectivity is not manifested by observers having the same sensory experience of it, but by their being able, through their diverse sensory experiences to acquire a common *understanding* of it.

I think this emphasis on intelligibility as the clue to reality is very much to the point.

The wavefunction is the vehicle of our understanding of the quantum world. Judged by the robust standards of classical physics it may seem a rather wraith-like entity. But it is certainly the object of quantum mechanical discourse and, for all the peculiarity of its collapse, its subtle essence may be the form that reality has to take on the atomic scale and below. Anyone who has had to teach a mathematically based subject will know the difficulties which students encounter in negotiating a new level of abstraction. They have met the idea of a vector as a crude arrow. You now explain to them that it is better thought of as an object with certain transformation properties under rotations. 'But what is it *really*?' they say. You implore them to believe that it is an object with certain transformation properties under rotations. They do not believe you; they think that you are holding back some secret clue that would make it all plain. Time and experience are great educators. A year later the student cannot conceive why he had such difficulty and suspicion about the nature of vectors. Perhaps we are in the midst of a similar, if much longer drawn out, process of education about the nature of quantum mechanical reality. If we are indeed in such a digestive, living-with-it, period it would explain something which is otherwise puzzling. A great many theoretical physicists would be prepared to express some unease about the conceptual foundations of quantum mechanics – in particular, about Copenhagen orthodoxy – but only a tiny fraction of them ever direct serious attention to such questions. Perhaps the majority are right to submit themselves to a period of subliminal absorption.

I do not pretend that these notions can do more than offer a way of thinking about the quantum world with some hope of doing justice both to the idiosyncracy of its ways and also to the beautiful structure of the microworld which has been laid bare by the *discoveries* of elementary particle physics.

Appendix

Some mathematical knowledge, including an elementary grasp of the calculus, adds considerably to one's ability to understand the details of quantum theory. The following paragraphs present additional material accessible to readers with such a background.

A1 The Bohr atom

Consider an electron of charge $-e$ and mass m moving in a circle around a proton of charge e which is sufficiently massive to be treated as fixed. (The mass of the proton is 1836 times the mass of the electron so that this is a good approximation.) Let the radius of the circle be r and the velocity of the electron v. The electrostatic attraction between the electron and the proton must exactly match the electron's mass times its centrifugal acceleration in the circular orbit:

$$\frac{e^2}{r^2} = m\frac{v^2}{r}, \qquad\qquad [A1.1a]$$

or

$$e^2 = mv^2 r. \qquad\qquad [A1.1b]$$

The energy of the electron is made up of its kinetic and electrostatic potential energies:

$$E = \frac{1}{2}mv^2 - \frac{e^2}{r}, \qquad\qquad [A1.2]$$

which, by [A1.1b], reduces to

$$E = -\frac{e^2}{2r}. \qquad\qquad [A1.3]$$

The Bohr condition for the quantisation of the angular momentum (mvr) is

$$mvr = n\hbar, \quad n = 1, 2, \ldots. \quad [A1.4]$$

[A1.1b] and [A1.4] together give

$$r = n^2\hbar^2/me^2. \quad [A1.5]$$

Thus [A1.3] evaluated subject to [A1.5] yields.

$$E_n = \frac{-e^4m}{2\hbar^2} \cdot \frac{1}{n^2}, \quad [A1.6]$$

which is just the Bohr formula for the energy levels of the hydrogen atom, leading to the Balmer series by the argument of Chapter 2.

A2 De Broglie waves

In the formalism of special relativity space-and-time and momentum-and-energy form natural combinations. (Technically, they are four-vectors.) The Planck formula

$$E = h\nu \quad [A2.1]$$

makes *energy* proportional to frequency which is just the number of vibrations per unit interval of *time*. To preserve relativistic symmetry, de Broglie proposed that *momentum* should be similarly proportional to the inverse of wavelength, that is to the number of vibrations per unit interval of *space*. Thus he wrote

$$p = h/\lambda, \quad [A2.2]$$

where p is the momentum, λ is the wavelength and h is again Planck's constant. The equations [A2.1] and [A2.2] give a natural way to associate a wave (characterised by ν and λ) with a particle (characterised by E and p).

A3 The Schrödinger equation

A travelling wave moving in the x direction and specified by [A2.1, 2] corresponds to the wavefunction

$$\psi(x, t) = e^{i(kx - \omega t)}, \quad [A3.1]$$

where the angular frequency ω is given by

$$\omega = 2\pi v, \quad\quad\quad\quad\quad\quad\quad [\text{A3.2}]$$

and the wavenumber k is defined by

$$k = 2\pi/\lambda. \quad\quad\quad\quad\quad\quad\quad [\text{A3.3}]$$

Since

$$i\hbar \frac{\partial \psi}{\partial t} = E\psi, \quad\quad\quad\quad\quad\quad\quad [\text{A3.4}]$$

$$-i\hbar \frac{\partial \psi}{\partial x} = p\psi, \quad\quad\quad\quad\quad\quad\quad [\text{A3.5}]$$

the energy and momentum associated with the wave are the eigen-values of the differential operators $i\hbar\partial/\partial t$ and $-i\hbar\partial/\partial x$, respectively. This suggests, according to the ideas of Chapter 3, that these operators should correspond to energy and momentum in more general circumstances, so that we identify

$$E \text{ with } i\hbar \frac{\partial}{\partial t}, \quad\quad\quad\quad\quad\quad\quad [\text{A3.6}]$$

$$p_x \text{ with } -i\hbar \frac{\partial}{\partial x}, \quad\quad\quad\quad\quad\quad\quad [\text{A3.7a}]$$

or, generalising the latter to three spatial dimensions,

$$p \text{ with } -i\hbar \nabla, \quad\quad\quad\quad\quad\quad\quad [\text{A3.7b}]$$

where ∇ is the vector gradient operator

$$\nabla = \left(\frac{\partial}{\partial x}, \frac{\partial}{\partial y}, \frac{\partial}{\partial z} \right). \quad\quad\quad\quad\quad\quad\quad [\text{A3.8}]$$

The classical energy E is the sum of the kinetic and potential energies. For a particle of mass m moving in a potential V this gives

$$E = \frac{p^2}{2m} + V \quad\quad\quad\quad\quad\quad\quad [\text{A3.9}]$$

Translating [A3.9] into terms of the quantum mechanical operators (A3.6, 7b) and allowing the differential operators to act on the wavefunction $\Psi(x, t)$ one obtains the Schrödinger equation

$$ i\hbar \, \frac{\partial \Psi}{\partial t} = \left[- \frac{\hbar^2 \nabla^2}{2m} + V(x) \right] \Psi. \qquad [A3.10] $$

The operator in square brackets is called the Hamiltonian.

A4 Linear spaces

Vectors $| \alpha_i \rangle$ form a linear vector space if any linear combination of them,

$$ \sum_i \lambda_i \, | \alpha_i \rangle, \qquad [A4.1] $$

also belongs to the space. For quantum mechanical applications the coefficients λ_i are allowed to be complex numbers.

The dual space of bra vectors is antilinearly related to the ket vector space:

$$ \underset{\text{ket}}{\sum_i \lambda_i \, | \alpha_i \rangle} \rightarrow \underset{\text{bra}}{\sum_i \langle \alpha_i \, | \, \lambda_i^*}, \qquad [A4.2] $$

where * represents complex conjugation. In the concrete example of wave mechanics the kets $| \alpha_i \rangle$ are wavefunctions ψ_i and the bras $\langle \alpha_i |$ are the complex conjugate wavefunctions ψ_i^*.

A scalar product

$$ \langle \alpha' \, | \, \alpha \rangle \qquad [A4.3] $$

is defined between any ket $| \alpha \rangle$ and any bra $\langle \alpha' |$. It is a complex number with the property that

$$ \langle \alpha \, | \, \alpha' \rangle = \langle \alpha' \, | \, \alpha \rangle^*. \qquad [A4.4] $$

A consequence of [A4.4] is that $\langle \alpha \, | \, \alpha \rangle$ is real. We also require that it is positive,

$$ \langle \alpha \, | \, \alpha \rangle > 0. \qquad [A4.5] $$

The idea is that it is the square of the length of the vector $| \alpha \rangle$.

In concrete wave mechanical terms the scalar product corresponds to the integral

$$ \int \psi'^* \psi \, dx, \qquad [A4.6] $$

which obviously has the property [A4.4], just as

$$ \int \psi^* \psi \, dx \qquad [A4.7] $$

has the property [A4.5], since $\psi^* \psi = | \psi |^2$ is positive.

The correspondence between kets and physical states is what is called a ray representation. This means that $|\alpha\rangle$ and $\lambda|\alpha\rangle$ correspond to the same physical state, where λ is any non-zero complex number.

A5 Hermitean operators

Operators can be made to look like matrices by inserting them into a sandwich between a bra and a ket to give

$$\langle \alpha \mid A \mid \beta \rangle, \qquad [A5.1]$$

which is the analogue of a matrix element. The hermitean conjugate $A^\dagger$ of an operator A is defined by its matrix elements satisfying

$$\langle \alpha \mid A^\dagger \mid \beta \rangle = \langle \beta \mid A \mid \alpha \rangle^*. \qquad [A5.2]$$

That is, they are the transposed complex conjugates of the matrix elements of A, just as is the case for ordinary matrix hermitean conjugation.

An operator which is its own hermitean conjugate,

$$A^\dagger = A, \qquad [A5.3]$$

is said to be *hermitean*. Suppose $|a'\rangle$ is an eigenket of such an A with eigenvalue a':

$$A \mid a' \rangle = a' \mid a' \rangle. \qquad [A5.4]$$

Multiplying by the bra $\langle a' \mid$ gives

$$\langle a' \mid A \mid a' \rangle = a' \langle a' \mid a' \rangle. \qquad [A5.5]$$

(The number a', not being an operator, is not caught in the bra–ket sandwich.) From [A5.2, 3] it follows that $\langle a' \mid A \mid a' \rangle$ is real and we already know that $\langle a' \mid a' \rangle$ is real and non-negative (see [A4.5]). Thus we deduce that a' is real. This is the important property, stated in Chapter 3, that the eigenvalues of hermitean operators are real.

The eigenkets of observables form what the mathematicians call a complete set. This means that any other ket can be expanded in terms of them:

$$\mid \alpha \rangle = \sum_i \lambda_i \mid a_i' \rangle. \qquad [A5.6]$$

In this equation the arbitrary ket $\mid \alpha \rangle$ is shown expanded in terms of the eigenkets $| a_i \rangle$ of the observable A, the coefficients in the expansion being λ_i. If $| \alpha \rangle$ and the $| a_i \rangle$ are all normalised, that is if they satisfy

$$\langle \alpha \mid \alpha \rangle = 1, \langle a_i \mid a_i \rangle = 1, \qquad [A5.7]$$

then the λ_i are what are called *probability amplitudes*. It turns out that the probability of measuring A in the state represented by $| \alpha \rangle$ and finding the answer a is given by a sum of the squares of moduli of probability amplitudes:

$$\sum_{a_i = a} | \lambda_i |^2, \qquad [A5.8]$$

the sum being taken only over those λ_i whose corresponding eigenvalue a_i' is equal to the specified result a.

A6   Heisenberg and Schrödinger pictures

Schrödinger represents the dynamical development in time of a ket by his equation

$$i\hbar \frac{\partial}{\partial t} | \alpha, t \rangle = H | \alpha, t \rangle. \qquad [A6.1]$$

For simplicity let us suppose, as is frequently the case, that the Hamiltonian H does not depend explicitly on the time. Then [A6.1] can be integrated to give

$$| a, t \rangle = e^{-iHt/\hbar} | \alpha, 0 \rangle. \qquad [A6.2]$$

By hermitean conjugation the corresponding equation for a bra is

$$\langle \alpha, t | = \langle \alpha, 0 | e^{iHt/\hbar}. \qquad [A6.3]$$

The physical interpretation of the theory is all expressed in terms of bra−operator−ket sandwiches of the form

$$\langle \alpha | A | \beta \rangle. \qquad [A6.4]$$

In the Schrödinger picture the time dependence of [A6.4] is entirely due to the time dependence of the bra and ket. By [A6.2, 3] this can be written

$$\langle \alpha, 0 | e^{iHt/\hbar} \cdot A \cdot e^{-iHt/\hbar} | \beta, 0 \rangle. \qquad [A6.5]$$

Bracketing the factors differently, this can also be written in the form

$$\langle \alpha, 0 \mid A(t) \mid \beta, 0 \rangle.$$ [A6.6]

Here

$$A(t) = e^{iHt/\hbar} A e^{-iHt/\hbar}.$$ [A6.7]

(Since, in general A and H do not commute the two exponential factors do not simply cancel.) All the time dependence is now concentrated in the operator $A(t)$. This is just the Heisenberg dynamical picture. The equation of motion for A is easily found to be

$$i\hbar \frac{dA(t)}{dt} = A(t)H - HA(t) \equiv [A(t), H].$$ [A6.8]

A7 Identical particles

Suppose we have a wavefunction $\psi(x_1, x_2)$ corresponding to two identical particles with co-ordinates x_1 and x_2. Then, because of the identity of the particles, $\psi(x_2, x_1)$ must be the same physical state. By the idea of a ray representation of states in terms of wavefunctions (see A4), this means that

$$\psi(x_2, x_1) = \lambda \, \psi(x_1, x_2),$$ [A7.1]

where λ is some complex number. Now let us interchange 1 and 2 a second time. Double application of [A7.1] gives

$$\lambda^2 \, \psi(x_1, x_2).$$ [A7.2]

On the other hand, two interchanges must leave us exactly where we started, that is back with $\psi(x_1, x_2)$. From this we deduce that

$$\lambda^2 = 1, \text{ or } \lambda = \pm 1.$$ [A7.3]

The plus sign gives Bose statistics, the minus sign Fermi statistics. [Technical note of caution: this argument does not extend to systems of three or more particles in quite as straightforward a manner as one (or many books) might suppose.]

A8 The γ-ray microscope

The resolving power of a microscope of angular aperture 2α and using light of wavelength λ gives an uncertainty in image position

Δx, where

$$\Delta x \sim \frac{\lambda}{\sin \alpha}, \qquad\qquad [A8.1]$$

the sign $\sim$ indicating an order of magnitude estimate.

The uncertainty in the photon's momentum in a direction transverse to the axis of the microscope (which, by momentum balance, is also the uncertainty in the momentum transferred to the electron) is given by

$$\Delta p \sim 2p \sin \alpha \sim \frac{2h \sin \alpha}{\lambda}. \qquad\qquad [A8.2]$$

combining [A8.1] with [A8.2] gives

$$\Delta x \cdot \Delta p \sim 2h \geq h, \qquad\qquad [A8.3]$$

in accordance with Heisenberg's uncertainty relation.

Notice the important role of the factor $\sin \alpha$ which relates to the angular aperture of the microscope (see the discussion of p. 48).

A9 The Bell inequality

Let $N(+\ +\ +)$ be the number of particles in the test with α_+, β_+, γ_+; etc. Let $N(\alpha_+\ \beta_+)$ be the number of particles with α_+, β_+ and γ unspecified; etc. Then

$$\begin{aligned}
N(\alpha_+\ \beta_-) &= N(+\ -\ +) + N(+\ -\ -), \\
N(\alpha_+\ \gamma_-) &= N(+\ +\ -) + N(+\ -\ -), \\
N(\beta_-\ \gamma_+) &= N(+\ -\ +) + N(-\ -\ +).
\end{aligned} \qquad [A9.1]$$

Since all the Ns are non-negative, it follows that

$$N(\alpha_+\ \beta_-) \leqslant N(\alpha_+\ \gamma_-) + N(\beta_-\ \gamma_+). \qquad [A9.2]$$

Recall that when one particle has α_+ then the other particle of the pair must have α_-; etc. This implies that the quantities n defined in Chapter 7 are proportional to sums of pairs of Ns, according to the scheme:

$$\frac{n(\alpha_+\ \beta_+)}{N(\alpha_+\ \beta_-)+N(\alpha_-\ \beta_+)} = \frac{n(\alpha_+\ \gamma_+)}{N(\alpha_+\ \gamma_-)+N(\alpha_-\ \gamma_+)} =$$

$$\frac{n(\beta_+\ \gamma_+)}{N(\beta_+\ \gamma_-)+N(\beta_-\ \gamma_+).} \qquad [A9.3]$$

From [A9.2] and the corresponding inequality with $+$, $-$, interchanged, it follows that

$$n(\alpha_+\ \beta_+) \leqslant n(\alpha_+\ \gamma_+) + n(\beta_+\ \gamma_+), \qquad [A9.4]$$

which is the Bell inequality of Chapter 7.

Glossary

Inevitably definitions interlock. Terms that are treated in this glossary are printed in *italic* when they are used elsewhere as part of another definition.

action An important dynamical quantity, related to the 'business' of a system, which can be measured in units of *Planck's constant*.

α-particle The *nucleus* of a helium atom.

angular momentum A dynamical quantity which measures the amount of rotatory motion present in a system. It can be measured in units of *Planck's constant*.

Balmer formula A simple formula (p. 9) which gives the frequencies of certain lines prominent in the spectrum of hydrogen.

Bell inequality An inequality (p. 76) satisfied by certain *spin* correlations in *locally realistic* theories. It does not hold in quantum mechanics nor, it seems, in nature.

black body radiation Radiation in equilibrium with the walls of a container which is perfectly absorbing and re-emitting.

Bohr atom An account of the hydrogen atom which was a contrived but instructive half-way house between *classical physics* and quantum theory.

bosons Particles whose *statistics* require them to be in *states* corresponding to symmetric *wavefunctions*.

bra vector See *ket vector*.

Brownian motion The random joggling motion of light particles suspended in a fluid.

classical physics Strictly deterministic physics constructed according to the laws of Newton and Maxwell. It provides a good description of the behaviour of systems whose *action* is large compared with *Planck's constant*.

collapse of the wave packet The instantaneous change in the *wavefunction* consequent on making a *measurement* of an *observable*. The wavefunction 'collapses' onto the *eigenvector* corresponding to the *eigenvalue* which is the actual result of the measurement.

commutativity The property of multiplication being independent of the order of the factors; that is, $a \times b = b \times a$. Not all mathematical forms of multiplication possess this property, in which case they are said to be non-commutative.

complementarity The characteristic quantum mechanical fact that alternative and mutually exclusive descriptions are possible for dynamical systems. For example, one can know where a particle is (a description in terms of position) in which case the *uncertainty principle* does not permit one to know what it is doing, or one can know what it is doing (a description in terms of momentum) in which case one does not know where it is.

complex numbers A system of numbers including i, the square root of -1.

Copenhagen interpretation The interpretation of the *measurement* process in quantum theory, espoused by Bohr and his colleagues, which assigns the reason for obtaining a particular result on a particular occasion to the effect of the intervention of classical measuring instruments. According to this point of view the relevant entity to consider is always (system observed plus measuring instruments), so that a change in the disposition of the measuring instruments, even without any change in the system observed, is considered to create a totally new situation.

correspondence principle The requirement that the results of *classical physics* must be recoverable from quantum theory as a good approximation for the behaviour of systems whose *action* is large compared with *Planck's constant*.

differential equations Mathematical equations which govern the evolution of a system in an orderly and continuous fashion.

eigenvector A *vector* which when operated upon by a specified *linear operator* reproduces the original vector multiplied by a numerical factor. This factor is called the corresponding *eigenvalue*.

Einstein–Podolsky–Rosen paradox The surprising consequence of quantum theory that once two systems have interacted with each other, then a measurement on one system can produce an instantaneous change in the *state* of the other system, even if they are by then widely separated from each other.

electron An elementary particle which together with the *nucleus* forms the constituents of atoms.

electron diffraction The fact that *electrons* manifest the phenomenon of *interference* when passed through thin metal films. This demonstrates the wave-like character of electrons.

exclusion principle The requirement that two *electrons* (more generally, two *fermions*) cannot occupy the same *state of motion*.

fermions Particles whose *statistics* require them to be in *states* correspond-to antisymmetric *wavefunctions*.

Fourier analysis The mathematical theory which expresses quantities in terms of the addition of waves of different frequencies.

γ-ray microscope An important *thought experiment* used by Heisenberg to test the idea of the *uncertainty principle*. It endeavours to locate a particle as accurately as possible, using a microscope with short-wavelength illumination, whilst at the same time minimising the degree of uncontrollable disturbance to which the particle is subjected.

general relativity The modern theory of gravity which expresses the effects of gravitational forces on particles in geometrical terms of the curvature of space-time.

Hamiltonian The *observable* associated with the energy of a system.

hermitean A mathematical property of certain *linear operators* which ensures that their *eigenvalues* are all real.

hidden variable theories Attempts to reconcile the randomness of the quantum measurement process with an underlying strictly determinate dynamical picture. Such theories associate the need for a *probability interpretation* with our ignorance of certain ('hidden') aspects of the system studied.

Hilbert space A mathematical theory of *vectors* in a space of infinite dimensions.

idealism The philosophical standpoint that assigns reality only to mental phenomena.

interference A characteristic phenomenon of wave motion in which two waves interact with each other, at some points reinforcing each other, at other points cancelling each other out. The result is an interference pattern composed of alternations of large and small effects.

ket vector The *vector* which represents a *state of motion* in quantum theory. Each ket has associated with it a *bra vector* (called by the mathematicians its dual) which represents the same state.

Lamb shift A small frequency difference between two adjoining lines in the hydrogen spectrum.

level problems The question of how different levels of description of the world (physics, biology, psychology etc.) relate to each other.

linear operator A mathematical entity turning one *vector* into another and satisfying certain (linear) conditions for its action upon sums and multiples (p. 26).

locality The requirement that causes and their immediate effects occur at the same place (no action-at-a-distance).

locally realistic theory One that combines *realism* with *locality*.

many-worlds interpretation The interpretation of quantum *measurement* which asserts that at the moment of each act of measurement the universe splits up into many parallel and disconnected universes, in each of which one of the possible results of the measurement actually occurs.

matrix mechanics The version of quantum theory invented by Heisenberg. It is close in its formalism to that of *classical physics* but it requires the *observables* to be represented by (non-commuting) matrices.

measurement The act of ascertaining the value taken by an *observable* of a given system on a particular occasion.

non-locality The property of permitting a cause at one place to produce immediate effects at distant places.

nucleus The positively charged central core of an atom in which almost all the matter is concentrated. Nuclei are composed of protons and neutrons.

observable A quantity associated with a physical system which can be measured experimentally.

particle–wave duality The fact that entities such as *electrons* and *photons* behave both as if they were particles and also as if they were waves.

path-integral formalism A way of formulating quantum theory which pictures the evolution of a system as due to a superposition of all possible trajectories. For that reason it is sometimes called the *sum over histories*.

photoelectric effect The phenomenon in which *electrons* are ejected from a metal by the incidence of a beam of light.

photon The particle of light.

pilot wave A *hidden variable* version of quantum theory in which the wave acts as guide for the motion of its associated particle.

Planck's constant The fundamental unit of *action* ($h = 6.63 \times 10^{-34}$ joule-seconds) which sets the scale for 'large' and 'small' in quantum theory.

polarisation The direction of oscillation of a wave. Light waves are transversely polarised, that is they oscillate in directions perpendicular to their direction of motion.

positivism The philosophical standpoint that asserts that the quantities of science are solely those which are directly measurable and that the role of science is harmoniously to reconcile the results of measurement without postulating an underlying reality which is their source.

probability amplitude A complex number, the square of whose modulus gives a probability.

probability interpretation The interpretation of the *wavefunction* of *wave mechanics* as being the *probability amplitude* for calculating the probability of finding a particle at a particular point. More generally, the notion that quantum theory can only calculate the relative probabilities of obtaining certain results from the *measurement* of an *observable* and is unable to predict which specific result will be obtained on a particular occasion of measurement.

quantum electrodynamics The theory of the interaction of *electrons* and *photons*.

quantum field theory The result of applying quantum theory to the behaviour of a field, such as the electromagnetic field. The resulting formalism resolves the apparent paradox of *particle–wave duality*.

quantum logic The modification of the distributive laws of logic required by the existence of new possibilities permitted by the *superposition principle* of quantum theory.

quarks and gluons The currently accepted fundamental constituents of matter out of which particles like protons and neutrons are composed.

realism The philosophical standpoint which asserts the reality of a world external to the observer, whose detailed structure is considered to be open to the investigations of science.

Schrödinger equation The fundamental *differential equation* governing the evolution of a physical system in *wave mechanics*.

spectroscopy The study of the line structure found in the light emitted by heated gases when the light is decomposed by a prism into its component frequencies.

spin The intrinsic *angular momentum* possessed by particles like *electrons* and *photons*.

state of motion of a system is one in which as much as possible is specified about the details of the motion.

statistics The behaviour of systems of several identical particles, in particular the symmetry or antisymmetry of their *wavefunction* under the interchange of a pair of particles.

Stern–Gerlach experiment An experiment to measure the *spin* of a particle by observing its deflection by an inhomogeneous magnetic field.

Superposition principle The quantum mechanical principle which permits the formation of *states* of a system by the superposition, or adding together, of other states. Such combinations are to be interpreted in a probabilistic way, so that the state formed by the superposition has certain probabilities for exhibiting the properties of the states out of which it is composed.

sum over histories See *path integral formalism*.

thought experiments Idealised experiments conceived as tests in principle of the consistency or scope of a physical theory.

tunnelling The intrinsically quantum mechanical phenomenon by which particles can penetrate through regions that they would have insufficient energy to enter in *classical physics*.

ultraviolet catastrophe The result from *classical physics* which, if correct, would assign infinite energy to the high frequencies of *black body radiation*.

uncertainty principle The quantum mechanical result, discovered by Heisenberg, which asserts that certain pairs of *observables* (such as position and momentum, or time and energy) cannot both be known to a greater degree of accuracy than is specified by a limit expressed in terms of *Planck's constant* (p. 45).

vectors Mathematical quantities which can be added together and multiplied by numerical factors.

wavefunction The quantity which represents the *state* of a physical system in *wave mechanics*. It is a particular example of a *ket vector*.

wave mechanics The version of quantum theory, invented by Schrödinger, which emphasises the wave-like character of physical systems.

Further reading

The following is a personal selection of books which may be found useful by those who wish to pursue these matters further:

A book which surveys many of the topics of the present volume, at an approximately similar level but in a very different style and with differing emphases is

Paul Davies: *Other Worlds* (Dent, 1980).

Two books which survey modern physics and seek to assimilate it to Eastern thought are

Fritjof Capra: *The Tao of Physics* (Fontana, 1976),

Gary Zukav: *The Dancing Wu Li Masters* (Fontana, 1980).

Personally I feel that the attempt depends too greatly on purely verbal parallels to be convincing.

Those who wish to see how quantum ideas are used in the modern theory of the structure of matter could read

J.C.Polkinghorne: *The Particle Play* (W.A. Freeman, 1979),

especially chapters V and VII which contain material on quantum field theory.

An account of the development of quantum mechanics is given in

M. Jammer: *The Conceptual Development of Quantum Mechanics* (McGraw-Hill, 1966).

Less satisfactory overall, to my mind, but nevertheless containing some interesting material is

M. Jammer: *The Philosophy of Quantum Mechanics* (John Wiley, 1974).

A fascinating account of the development of Einstein's thought, including his attitude to quantum mechanics, is to be found in

Abraham Pais: *'Subtle is the Lord ...'* (Oxford University Press, 1982),

a splendid scientific biography of a very great man.

Among the more interesting and accessible of the general writings of the Founding Fathers are:

Niels Bohr: *Atomic Physics and Human Knowledge* (John Wiley, 1958),

which contains an account of Bohr's long struggle with Einstein over the uncertainty principle, illustrated by Bohr's charming naive drawings of the 'apparatus' employed in the thought experiments they discussed, and

Werner Heisenberg: *Physics and Philosophy* (Allen and Unwin, 1959). The classic account of the basic principles of quantum mechanical formalism is

P.A.M. Dirac: *The Principles of Quantum Mechanics* (Oxford University Press, 4th edition, 1982).

There are many good textbooks from which the mathematical physicist can learn the fundamental calculational techniques. A typical example is

L. Schiff: *Quantum Mechanics* (McGraw-Hill, 3rd edition, 1968).

However most books which give accounts of the formalism are strangely reluctant to discuss the sort of interpretative problems which have been our main concern. For a professional account of these questions one would best start with

B. d'Espagnat: *Conceptual Foundations of Quantum Mechanics* (Benjamin, 1971).

Index